Enclosed Experimental Ecosystems and Scale:

Tools for Understanding and Managing Coastal Ecosystems

Enclosed Experimental Ecosystems and Scale:

Tools for Understanding and Managing Coastal Ecosystems

Edited by John E. Petersen, Victor S. Kennedy, William C. Dennison, and W. Michael Kemp

Authors:

G. Mine Berg	W. Michael Kemp
Maureen Brooks	Victor S. Kennedy
Chung-Chi Chen	Rober P. Mason
Jefferey C. Cornwell	Laura Murray
William C. Dennison	Roger I.E. Newell
Robert H. Gardner	John E. Petersen
Pat M. Glibert	Elka T. Porter
Tim Goertemiller	Lawrence P. Sanford
Debroah C. Hinkle	J. Court Stevenson
Edward D. Houde	Karren L. Sundberg
Pat Kangas	Steven E. Suttles

Science Communication by Tracey A. Saxby and Joanna L. Woerner

 Springer

Editors
John E. Petersen
Lewis Center for Environmental Studies
Oberlin College
Oberlin, OH, US

Victor S. Kennedy
Horn Point Laboratory
University of Maryland Center for Environmental Science,
Cambridge, MD, USA

William C. Dennison
Integration and Application Network
University of Maryland Center for Environmental Science,
Cambridge, MD, USA

and

W. Michael Kemp
Horn Point Laboratory
University of Maryland Center for Environmental Science,
Cambridge, MD, USA

ISBN 978-0-387-76766-6 (Hard cover)   e-ISBN 978-0-387-76767-3
e-ISBN 978-0-387-76768-0 (Soft cover)
DOI: 10.1007/978-0-387-76767-3

Library of Congress Control Number: 2008940438

Printed on acid-free paper

springer.com

Preface

The wide range of trends and environmental challenges now facing human society, from population growth, to urbanization, to industrialization and modern agriculture, to species invasions and losses, to global climate change, are nowhere more evident than in the shallow waters that form the interface between land and sea. The development of sound environmental policy that effectively preserves and restores critical coastal resources is predicated on the acquisition of new scientific knowledge that elucidates the numerous interacting factors that control these complex systems. Research in the coastal zone is complicated by interactions that occur over broad scales of time, space, and ecological complexity, and by unique conditions that render controlled field research especially challenging in these environments. Enclosed experimental ecosystems (mesocosms and microcosms) have gained in popularity as research tools in coastal aquatic ecosystems in part because they provide scientists with a degree of experimental control that is not achievable through field experiments. Yet to date, techniques for systematically extrapolating results from small-scale experimental ecosystems to larger, deeper, more open, more biodiverse, and more heterogeneous ecosystems in nature have not been well developed. Likewise, researchers have lacked methods for comparing and extrapolating information among natural ecosystems that differ in scale. Experimental ecosystems represent a potentially powerful tool for testing and expanding our understanding of the mechanisms that drive ecological dynamics in the coastal zone. The information contained within the pages of this book is intended to help practitioners make the most of this promising approach to ecological research.

The lessons described in this book are drawn largely from a series of investigations conducted under the auspices of the Multiscale Experimental Ecosystem Research Center (MEERC). For a decade, the Environmental Protection Agency supported this comprehensive research effort within the University of Maryland Center for Environmental Science. MEERC researchers have intentionally manipulated and observed the effects of time, space, and ecological complexity on the dynamics of the range of aquatic habitats that dominate in the mid-Atlantic region of the United States. The mesocosms have been constructed to simulate benthic-pelagic, submersed aquatic vegetation, and marsh habitats. This book draws on the entire body of research that has been conducted in enclosed experimental ecosystems and distills the essence of what has been learned about experimental design and scale. Diagrams, conceptual models, and straight-forward language are emphasized so as to present information in a form that can be easily applied by scientists, students, managers, and policymakers.

This book has a range of intended audiences. Scientists who produce or interpret experimental ecosystem research will find it useful. The book provides principles for design and protocols for performing experimental ecosystem experiments. Resource managers and policymakers can also use this book to aid in their understanding of how to extrapolate from research done in laboratories or mesocosms to intact ecosystems. This book can serve as a handbook for students studying coastal ecosystem processes, particularly with regard to improving their ability to design and interpret experiments.

This book has two main objectives. The first objective is to provide scientists, managers, and policymakers with an introduction to what has been termed the "problem of scale" (Levin 1992) as it relates to research in the coastal zone (Gardner et al. 2001; Grice and Reeve 1982; Newell 1988; Seuront and Strutton 2004). What do we know about the effects of time, space, and ecological complexity in coastal ecosystems and what do we need to know in order to better understand and manage these systems? The second objective is to present information that will allow for improved design and interpretation of enclosed experimental aquatic ecosystems. How can researchers design enclosed experimental ecosystems that more accurately model natural ecosystems? What do scientists, managers, and policymakers need to know in order to interpret, extrapolate, and apply findings from these experiments to nature?

This book is designed to be used in two distinct ways. First, the sections are organized and sequenced so as to provide a comprehensive introduction to issues of scale and experimental design in the construction, execution, and interpretation of aquatic mesocosm experiments. Second, each section, each sub-section, and, indeed, each page of the book is designed to serve as a stand-alone reference that can be used as a quick introduction to a particular topic. The table of contents and a comprehensive index have been carefully constructed to guide the reader to desired content.

This book begins with a general introduction to the problem of scale and the role of experimentation and enclosed experimental ecosystems in coastal research. The next section outlines key design decisions that mesocosm researchers face and reviews lessons from multiscale experiments that should inform these decisions. Design decisions include such basic choices as container size, experimental duration, habitat type, and degree of biological, material, and energetic exchange. The third section explores tools available for designing and interpreting experimental ecosystems so as to obtain a desired balance between control and realism while maintaining statistical rigor. The book concludes with examples of management applications from a variety of experiments conducted in experimental aquatic ecosystems at the MEERC facility.

References

Gardner, R.H., W.M. Kemp, V.S. Kennedy, and J.E. Petersen. 2001. Scaling Relations in Experimental Ecology. Columbia University Press: New York.

Grice, G.D. and M.R. Reeve (eds). 1982. Marine mesocosms: Biological and chemical research in experimental ecosystems.Springer-Verlag: New York.

Levin, S.A. 1992. The problem of pattern and scale in ecology. Ecology 73: 1943-1967.

Newell R.I.E. 1988. Ecological changes in Chesapeake Bay: Are they the result of overharvesting the American oyster, *Crassostrea virginica?* Pages 536-546 in M.P. Lynch and E.C. Krome (eds.). Understanding the estuary: Advances in Chesapeake Bay research. Chesapeake Research Consortium Publication 129 (CBP/TRS 24/88), Gloucester Point, VA.

Seuront C. and P.G. Strutton. 2004. Handbook of scaling methods in aquatic ecology: Measurement, analysis, simulation. CRC Press, Boca Raton, FL.

Acknowledgements

The research described in this book was supported by the Environmental Protection Agency's STAR (Science to Achieve Results) program as part of the Multiscale Experimental Ecosystem Research Center (MEERC) at the University of Maryland Center for Environmental Science. The MEERC program built on lessons learned in other mesocosm facilities and benefited from the stimulation provided by a Science Advisory Committee (see table) composed of many of the leading thinkers on issues of scale and experimentation in coastal ecology.

Multiscale Experimental Ecosystem Research Center Science Advisory Committee

Member	Affiliation
Dr. Walter Adey	Smithsonian Institution Marine System Laboratory, Washington D.C.
Dr. Steve Bartell	Specialists in Energy, Nuclear, and Environmental Sciences Oak Ridge, Inc., Center for Risk Analysis, Oak Ridge, TN
Mr. Richard Batiuk	Environmental Protection Agency, Region III, Annapolis, MD
Dr. James Clark	Exxon Biomedical Sciences, Inc., East Millstone, NJ
Dr. Kenneth Dickson	Institute of Applied Sciences, University of North Texas, Denton, TX
Dr. James Kitchell	Laboratory of Limnology, University of Wisconsin, Madison, WI
Dr. Wayne Landis	Huxley College of Environmental Studies, WA
Dr. Raymond Lassiter	National Exposure Research Laboratory, EPA, Athens, GA
Dr. Simon Levin	Ecology and Evolutionary Biology, Princeton University, Princeton, NJ
Dr. Marlon Lewis	Oceanography, Dalhousie University, Halifax, Nova Scotia
Dr. Richard Lowrance	U.S.D.A., Agriculture Research Service, Tifton, GA
Dr. Irving Mendelssohn	Wetland Biogeochemistry Institute, Louisiana State University, Baton Rouge, LA
Dr. Robert Menzer	Gulf Breeze Environmental Research Laboratory, ORD, EPA, Gulf Breeze, FL
Dr. William Mitsch	School of Natural Resources, The Ohio State University, Columbus, OH
Dr. Hilary Neckles	U.S.G.S. Patuxent Wildlife Research Center, Augusta, ME
Dr. Scott Nixon	Graduate School of Oceanography, The University of Rhode Island, Narragansett, RI
Dr. Candace Oviatt	Graduate School of Oceanography, The University of Rhode Island , Narragansett, RI
Dr. Michael Pace	Institute of Ecosystem Studies, Millbrook, NY
Dr. Harry Pionke	U.S.D.A., Agriculture Research Service Northeast Watershed Research Center, University Park, PA
Dr. Tom Powell	Division of Environmental Studies, University of California, San Diego, CA
Dr. Don Scavia	National Oceanic and Atmospheric Administration, Coastal Ocean Office, Silver Springs, MD
Dr. David Schneider	Ocean Science Center, Memorial University of Newfoundland
Dr. John Steele	Woods Hole Oceanographic Institution, Woods Hole, MA
Dr. Frieda Taub	University of Washington, School of Fisheries, Seattle, WA

Numerous colleagues, students, and technicians contributed to this large research effort and to this book. A list of the principal investigators, technical staff, graduate students, post-doctoral fellows, and visiting scholars that assisted in MEERC research:

Multiscale Experimental Ecosystem Research Center Principal Investigators

Joel Baker	Patricia Glibert	Laura Murray
Brad Bebout	Jay Gooch	Roger Newell
Wayne Bell	Rodger Harvey	Jennifer Purcell
Walter Boynton	Edward Houde	Michael Roman
Russell Brinsfield	Todd Kana	Peter Sampou
Richard Calabrese	Michael Kemp	Lawrence Sanford
Daniel Conley	Christopher Madden	Kenneth Staver
Jeffrey Cornwell	Sharook Madon	Court Stevenson
Robert Costanza	Thomas Malone	Diane Stoecker
Hugh Ducklow	Robert Mason	Robert Ulanowicz
Thomas Fisher	Thomas Miller	
Robert Gardner	Raymond Morgan	

Technical Staff

Dan Fiscus	Sherry Pike	Brian Sturgis
Tim Goertemiller	John Posey	Steve Suttles
Deborah Hinkle	Alison Sanford	Tom Wazniak
Bryan Pearson	Heather Soulen	Jason Wyda
Elgin Perry	Karen Sundberg	Carmen Zarate

Graduate Students

Luis Abarca-Arenas	Angie Lawrence	Alison Sanford
Jeffrey Ashley	Joy Leaner	David Scheurer
Rick Bartleson	Amy Liebert	Josh Schmitz
Miné Berg	Xiping Ma	Kristin Schulte
John Brawley	Mike Mageau	William Severn
Maureen Brooks	Antonio Mannino	Fuh Kwo Shiah
Amy Chen	Kelly McAloon	Kellie Splain
Chung-Chi Chen	John Melton	Brian Sturgis
Teresa Coley	Jeff Merrell	Auja Sveinsdottir
Sean Crawford	Amy Merten	Stacey Swartwood
Joyce Dewar	William Mowitt	Arnold Turner
Chun-Lei Fan	Brandon Muffley	Lisa Wainger
Peter Hentschke	John Petersen	Jennifer Zelenke
Eun Hee Kim	Elka Porter	

Post-Doctoral Fellows

Steve Blumenshine	John Petersen
Elka Porter	Enrique Reyes

Visiting Scholars

Michael Heath, Marine Laboratory, Aberdeen	John Petersen, Oberlin College
Patrick Kangas, University of Maryland	Judith Stribling, Salisbury University

Contents

Introduction

J.E. Petersen, W.M. Kemp, V.S. Kennedy, W.C. Dennison, and P. Kangas

Coastal ecosystems are productive, economically important, and globally threatened by human population growth and climate change. Scientists and managers have come to realize that solutions to the burgeoning problems facing coastal ecosystems require a clear understanding of how the scales of time (from seconds to millennia), space (from the volume of a cell to the volume of the biosphere), and complexity (from a simple food chain to a complex food web) can affect our understanding of these ecosystems. In particular, how can findings from small-scale observations or experiments be applied to increase our knowledge of and ability to manage large natural ecosystems? This book is designed to help researchers, managers, and students answer this critical question (Cover Figure).

For over a century, scientists have used field surveys and field and laboratory experiments to investigate how coastal ecosystems function. For the last several decades, computer modeling studies have used data collected by such surveys and experiments to explore ecosystem processes and further advance our understanding.

In recent years, researchers have increasingly used enclosed experimental ecosystems (also called microcosms, mesocosms, enclosures, and limnocorrals) as tools to examine and manipulate ecosystems under replicated, controlled, and repeatable conditions in the field and in the laboratory. Well-designed enclosed systems can incorporate a diversity of biogeochemical processes and ecological interactions, making them living model ecosystems. Can results obtained from these model ecosystems be extrapolated to natural ecosystems? Can experiments in model ecosystems improve our understanding of the effects of scale on ecological processes and experimental observations?

These questions were the focus of research performed by scientists in the Multiscale Experimental Ecosystem Research Center (MEERC) within the University of Maryland Center for Environmental Science. Their findings form the foundation for this book. MEERC used mesocosms to simulate ecosystems in water columns and sediments, as well as salt marsh and aquatic grass habitats. The Center's objectives were to test scale-related theory, to improve experimental designs of mesocosms so as to retain key functions of nature, and to develop rules and tools for extrapolation from experiments to nature. Experiments were performed at population, community, and ecosystem levels of organization, with some experiments examining ecological responses to nutrients and contaminants.

This first section of the book presents enclosed experimental ecosystems as important research tools for exploring coastal ecology and introduces the key concepts and problems of scale that confront researchers, managers, and policy makers.

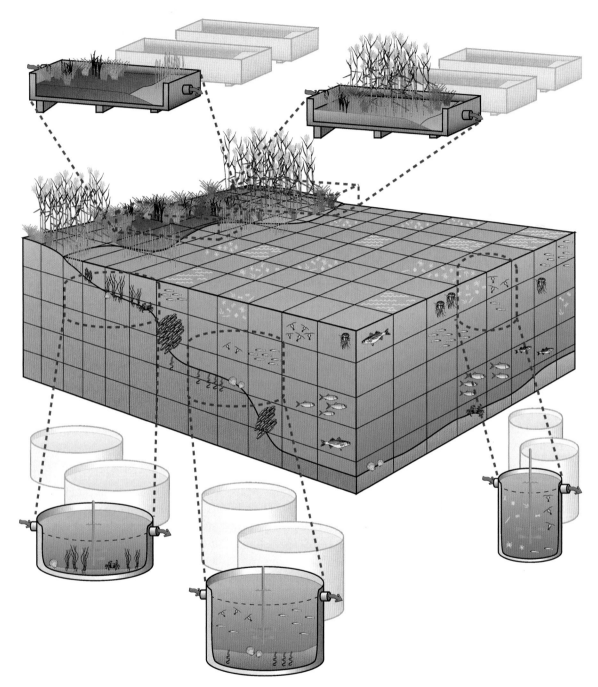

Cover Figure: *Enclosed experimental ecosystems (mesocosms) have become important tools that can be used to improve our understanding of and ability to manage coastal aquatic ecosystems. Mesocosms can be used to conduct controlled and replicated experiments that address a range of important questions related to the diverse and heterogeneous ecosystems that occupy the coastal zone. Scale is a critical consideration in mesocosm research. This book is designed to provide researchers and managers with a clear understanding of how to design experimental ecosystems and to how to extrapolate research findings to natural ecosystems.*

Coastal ecosystems, management, and research

The coastal zone is vital to humanity and is ecologically complex

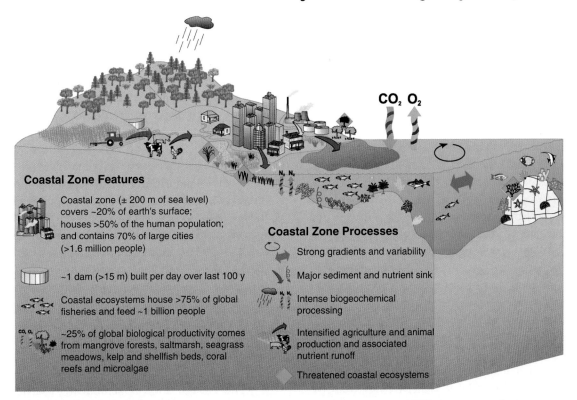

Figure 1: Overview of coastal zone features and processes (adapted from Dennison 2007). Human population pressure threatens many coastal ecosystems.

The coastal zone constitutes ~20% of the earth's surface (Fig.1). Over half of the global human population lives within this region, with this proportion growing due to population growth and migration to coastal regions. Seventy per cent of the world's megacities (> 1.6 million population) are in the coastal zone. The region is a nexus for commerce and other economic activities. Coastal waters support diverse communities such as mangrove forests, seagrass meadows, coral reefs, saltmarshes, rocky intertidal areas, and kelp and shellfish beds, as well as benthic and planktonic microalgae. Collectively, these ecosystems provide about 25% of global biological productivity and support more than 75% of global fisheries.

Physical, geological, biological, chemical, and human processes all shape the character of coastal ecosystems. Freshwater that enters from rivers, over land surfaces, and through groundwater transports water, soil, nutrients, and pollutants into coastal waters. Physical circulation, tides, and atmospheric forces transport materials between land and sea and produce nearshore gradients of salinity, sediments, and nutrients. Thus the coastal zone is an area of major biogeochemical activity, with elements such as carbon, nitrogen, phosphorus, iron, and silica undergoing important transformations. In the last century, rapid growth of human population accompanied by harvesting and species introductions, and by intensified agricultural, industrial, and residential activity with their associated pollution, nutrient, and sediment runoff have dramatically changed chemical and biological conditions in coastal waters.

Understanding the changes and the consequences of these changes is crucial to developing laws and management practices that conserve and restore these economically important ecosystems.

Major challenges to coastal ecosystems differ by region

There are marked differences in the major coastal issues affecting different parts of the world (Fig. 2). In temperate regions, the major issues are nutrient over-enrichment (eutrophication), overharvesting, and invasive species. Plant nutrients are released into coastal waters from point sources such as wastewater treatment and industrial facilities and from nonpoint sources such as agriculture activities (fertilizer application and animal manure). Nutrients released inland are delivered to the coastal zone through rivers, surface runoff, groundwater flow, and the atmosphere. Nutrients, particularly nitrogen and phosphorus, stimulate excess algal productivity. The algae shade seagrasses and deplete dissolved oxygen as they decompose. These changes in the plant community harm economically important fisheries and generally reduce biological diversity. Eutrophication can also increase the severity and abundance of toxic algal blooms.

Other critical issues for the temperate coastal zone include alterations in water flow that have resulted from land use changes, from extensive draining of wetlands, and from construction of dams, weirs, and irrigation systems.

In tropical regions, the major issues are coastal development, deforestation, global climate change, and aquaculture. Sediment and nutrient runoff from development and deforestation smothers organisms and hinders light transmission for photosynthesis. Aquaculture, and in particular shrimp impoundment, directly destroys mangrove forests as well as contributing to eutrophication. Coral reefs are among the most biologically diverse ecosystems in the world. Their rapid loss has been linked to eutrophication, development, warming of the ocean, and overharvesting.

In polar regions, the major issue is climate change. The polar coastal zone is typified by frozen land masses and sea ice. Atmospheric concentrations of carbon dioxide and other greenhouse gasses have steadily increased since the dawn of industrial activity, resulting in an unprecedented increase in global temperatures in the last few decades. Rapid warming in the polar regions has lead to melting of glaciers, pulsed runoff events in the Arctic, and reductions in sea ice thickness. Further warming has the potential to dramatically change ocean circulation patterns. In addition to climatic changes, toxicants that are transported from tropical and temperate regions to polar regions through the atmosphere and through mining and exploration have the potential to alter the polar coastal zone.

The coastal zone is critical to fisheries. Overharvesting of fish and shellfish both regionally and globally has lead to dramatic declines in fisheries. Even in the open ocean, there have been dramatic declines in abundance of top predators as a consequence of intense fishing pressures (Myers and Worm 2003).

Understanding both the cause and effect of coastal problems requires a diversity of research approaches conducted at a spectrum of scales.

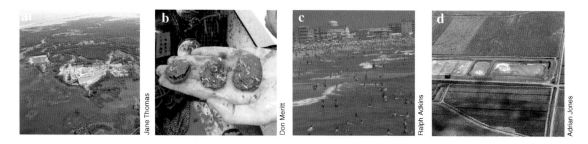

Figure 2: *Photos that depict the anthropogenic challenges in coastal systems: (**a**) Wetlands being filled for development in Maryland; (**b**) Declining oyster populations in Chesapeake Bay; (**c**) Crowded beach scene on the US east coast; and (**d**) Prawn (shrimp) aquaculture ponds surrounded by sugar cane fields in Australia.*

Scale is important in understanding coastal ecology

The field of ecology investigates interactions among organisms, and between organisms and their environments. Such interactions occur over scales that range in time from fractions of a second to millions of years and range in space from nanoliter volumes up to the volume of the entire biosphere (Fig. 3). Advances in technology, such as microscopy, mass-spectrometry, and remote sensing from satellites, allow ecologists to measure ecological changes over an unprecedented range of scales. However, even scientists tend to judge what is fast versus slow and what is small versus large in relation to scales associated with human perception and short-term human desires and needs.

Only within the last few decades have people begun to appreciate the fact that human activity affects ecological systems that range in scale from cells to the biosphere. Coastal ecologists, managers, and policymakers need to develop the ability to think about multiple scales.

A clear understanding of the scales of ecology is essential if humans are to address the environmental problems associated with local, regional, and global change. Effective communication among ecologists, among decision makers, and between these groups requires an improved understanding of how processes operate at different scales and a means of extrapolating information across scales. For example, how is the health of coastal waters simultaneously linked to coastal development, to management of individual farms, to emissions from cars in the airshed, and to climate change? How do researchers apply findings from small-scale experiments and measurements to large natural ecosystems? Answers to questions such as these are critical to the work of scientists, managers, and policy makers, and are also a necessary prerequisite to an informed public and to the development of rational environmental policy.

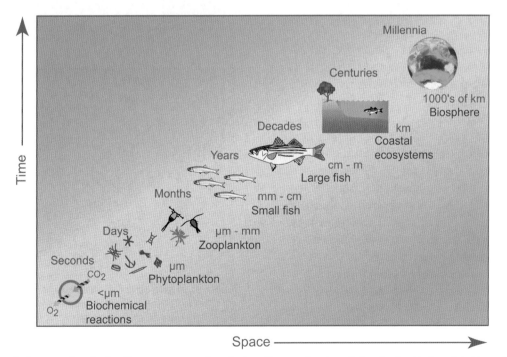

Figure 3: A range of scales is of interest to coastal ecologists. Temporal and spatial scales tend to be related to each other; small organisms or phenomena change relatively rapidly while large organisms or phenomena change more slowly.

Scale is inadequately addressed by managers and researchers

The effects of human activity are distinct at different scales (Fig. 4, top panel) and this has important implications for coastal management. For instance, conditions immediately offshore may be strongly influenced by a range of factors including wastewater and industrial outfalls, fertilizer applications, vegetative cover on immediately adjacent lands, and dredging activities in the local waterway. Over broader geographic regions, land-use management, water treatment, and industrial activities within all watersheds that empty into a coastal zone are key determinants of water quality and coastal resources. On yet broader scales of space and time, the character of coastal zones are influenced by shifts in temperature, precipitation, and water depth associated with human-induced global climate change.

To date, policy makers and managers have tended to focus their efforts on particular scales in isolation. The task of developing policy that simultaneously addresses the complete range of time and space scales over which human activities affect the coastal zone is challenging, but essential to effective management.

Ecologists also tend to emphasize particular scales of time and space in their research. For example, sub-disciplines in ecology range in focus from the study of environmental physiology of individual organisms to the study of energy flow and material cycling within whole ecosystems and even the entire biosphere. As a result, the questions, approaches, and analytical tools that scientists use to perceive the environment are distinct at these different scales (Fig. 4, bottom panel). Unfortunately, researchers working at a given scale often have a hard time understanding and communicating how their work relates to that of researchers working at other scales. For example, it is not always clear how mechanisms identified in small-scale, short-term, and simplified experimental systems apply to large, whole natural ecosystems. Too few researchers make the effort to consider scaling implications in analyzing their work.

Policymakers and managers will always be pressed to make decisions based on incomplete scientific understanding. The challenge for scientists and managers is to explicitly consider the effect of scale, and then to interpret and apply research findings within this context. New theory and new avenues of research are necessary to improve our ability to think and to interpret information across scales. This book distills and shares the recent developments in our understanding of scale.

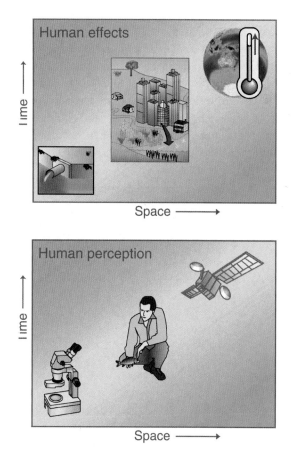

Figure 4: *Natural scales of time and space range from microscale to macroscale. Human effects on these ecological systems also occur at a various scales (top panel). In order to study these effects, a diversity of sensors are required to assist human perception at each scale (bottom panel).*

Experimental studies are increasingly emphasized in ecology

Ecology is still a relatively young science. Most scientific disciplines begin their history with an emphasis on observation and development of theory to explain these observations. In ecology this is reflected in the works of naturalists such as Charles Darwin and Ernst Haeckel. Observational studies, such as field surveys, are a critical means of developing hypotheses and predictions. However, the sign of a maturing science is an increased emphasis on experimentation as a means of testing the validity of these hypotheses. Within the last few decades there has been a clear trend within ecological science of growing reliance on manipulative experiments as a means of testing ecological theory (Fig. 5). Many approaches are available for experimentation.

One important distinction can be drawn between field and laboratory based experiments. In field experiments, either parts of nature or whole, naturally bounded ecosystems are manipulated in place while similar areas are left as controls. In laboratory experiments, organisms, communities, and the physical substrate are transported to controlled facilities for experimentation.

A second distinction can be drawn between experiments in which organisms and materials freely exchange between the experiment and a surrounding environment and those in which organisms and materials are enclosed and isolated either in a laboratory setting or with physical boundaries imposed in the field. The term *enclosed experimental ecosystem* is used when the goal of an enclosure experiment is to explore interactions among organisms or between organisms and their physical environment. Because enclosed experimental ecosystems are intended to serve as miniaturized worlds for studying ecological processes, they are often called microcosms or mesocosms (these terms will be used interchangeably). This book considers how enclosed experimental ecosystems can be used to understand and improve the management of coastal ecosystems, and focuses on the importance of scale in the design and interpretation of these experiments.

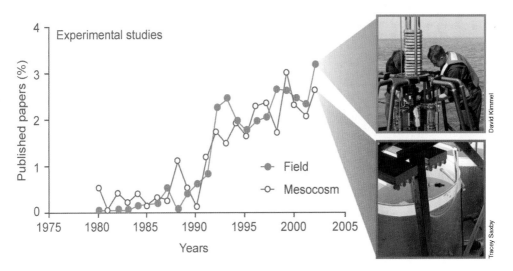

Figure 5: *Keyword searches were conducted in ecological journals to identify trends in the prevalence of field experiments and enclosed experimental ecosystems. These are not mutually exclusive categories; enclosed experimental ecosystems can be used in the field. The patterns suggest an increasing reliance on both categories of experimentation (Petersen et al. 2003).*

Various scientific approaches are needed to understand and manage coastal ecosystems

Field observations   Field experiments   Laboratory experiments   Simulation models

Add or subtract X   Control   +2X   +4X

Figure 6: *Scientists use field and laboratory observations and experiments to study coastal ecosystems. Models constructed from the resulting data can suggest new observations and experiments.*

All ecological systems can be divided into smaller subsystems. For example, the biosphere can be segregated into biomes, bioregions into watersheds, and ecosystems into living and non-living components. Scientists learn about ecological systems through research conducted at multiple scales in this hierarchy of subsystems (Fig. 6).

One powerful approach has been to study large complex systems by breaking them apart and examining their parts in isolation. For example, many chemical, physical, and biological processes that are important to ecosystem function (i.e., photosynthesis, nutrient uptake by plants, and simple predator–prey relations) can be studied using laboratory investigations that involve simplified chemical and physical systems

and include enzymes, individual cells, or one or a few species of organism. Such experiments are essential for advancing our understanding of individual processes and mechanisms controlling these processes.

However, the field of ecology recognizes that the whole is greater than the sum of its parts. Therefore, it is not possible to obtain a complete understanding of ecosystems by studying their parts in isolation; research on whole ecosystems is crucial to understanding the often complex dynamics that emerge when parts of the system interact with each other. The options available for ecosystem-level experimentation include manipulating whole ecosystems in nature or constructing enclosed experimental ecosystems in the field or laboratory (Fig. 7).

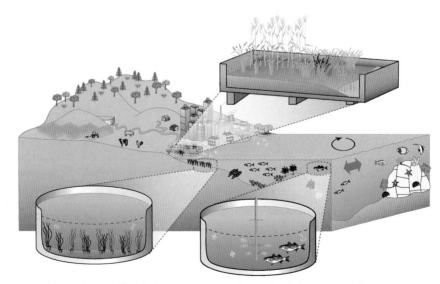

Figure 7: *The diverse interactions within whole ecosystems can be studied in intact natural ecosystems or at smaller scales in enclosed experimental ecosystems.*

Enclosed experimental ecosystems have become important tools for studying ecology at intermediate scales

Manipulating whole natural ecosystems poses enormous logistical challenges, is expensive, and in some cases is simply not feasible. Enclosed experimental ecosystems have become widely used tools in ecological research because they provide a means of conducting ecosystem-level experiments under replicated, controlled, and repeatable conditions at moderate expense. They provide an intermediate scale that can link information gained through small-scale laboratory experiments with information from field surveys (Fig. 8). Ecologists rely on both enclosed experimental ecosystems and field-based experiments (p. 7). In a quantitative review of the ecological literature, Ives et al. 1996 concluded that "our results show little difference between the publication and citation rates of microcosm experiments and other types of studies."

A principal difference between enclosed experimental ecosystems (mesocosms) and simple aquaria or terraria is that the former are designed to incorporate a more complete range of ecological interactions and biogeochemical processes; they are contained ecosystems rather than just organisms in a container. Because mesocosms are situated at an intermediate scale between simple laboratory studies and field studies, their use promises to advance our theoretical understanding of how ecosystems work and at the same time provide practical information that enables coastal systems to be managed more effectively.

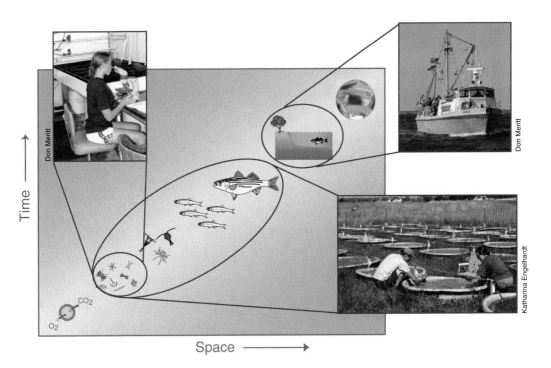

Figure 8: *Laboratory experiments tend to focus on short-lived organisms and small-scale processes. Field surveys are often used to study large-scale processes. Mesocosms, whether indoors or field based (as here), allow for study of ecosystems at intermediate scales of time and space.*

Experimental ecosystems complement other research approaches

Research approaches span a large range of scales (Table 1). At one end of this spectrum are field-based experiments including studies conducted on whole ecosystems, open plot experiments within natural ecosystems, and cage experiments in which certain organisms or processes are excluded from field plots while others are allowed to exchange freely across boundaries. At the other end of the spectrum are small-scale physiological experiments conducted in the laboratory, which make no attempt to simulate natural ecosystem processes.

Two fundamental objectives of ecological experiments are control and realism. *Control* refers to the degree and ease with which experimental conditions can be manipulated. *Realism* is a measure of the extent to which the experimental system and the experimental results accurately represent the dynamics of particular natural ecosystems. Trade-offs between control and realism are inherent in

different experimental approaches; experiments conducted within nature tend to maximize realism whereas physiological experiments in the laboratory allow for the highest degree of experimental control. Mesocosms provide intermediate levels of both control and realism (Fig. 9).

Mathematical simulation models are often used to synthesize and integrate findings from experiments conducted at different levels within the spectrum of research options. For instance, physiological experiments can be used to derive values for growth coefficients that are needed in ecosystem models. The results of mesocosm experiments and whole-ecosystem manipulations can then be used to calibrate and validate these models. Ultimately, calibrated models can be used to predict responses at dimensional scales and in response to ecological conditions that cannot be easily duplicated either in mesocosms or in field experiments.

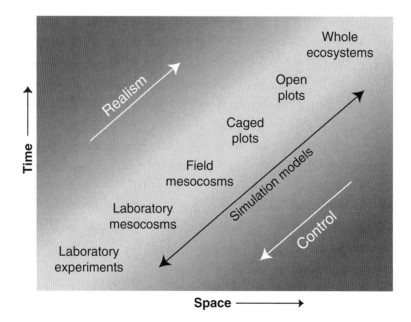

Figure 9: As the scale of experiments increases from simple laboratory experiments to complex whole ecosystem manipulations, greater realism results but control over experimental conditions declines. Simulation models can be used to synthesize and integrate results from the various types of empirical studies listed.

Table 1: Characteristics of different classes of experimentation.

Class of experiment	Scale	Methods	Advantages	Disadvantages	Examples
Whole ecosystem manipulations in nature	Scale of small natural ecosystems	Manipulate one or more factors. Maintain both control and treatment ecosystems.	High degree of realism.	Expensive; difficult to control and replicate for large ecosystems with environmental gradients.	Whole watershed manipulations: experimental lakes, tide pools, or small creeks.
In situ field experiments	$1\ m^2$–1 ha	Manipulate individual plots or water masses, each of which is open to the larger natural ecosystem.	High degree of realism and replication possible.	Treatment options constrained by openness. Patchy environments present challenges for replication.	Enrichment of Nutrients on a Coral Reef Experiment (ENCORE), Iron Enrichment Experiments (IEX).
Cage experiments	1–$100\ m^2$	Semipermeable enclosure allows water and small organisms to move in or out and excludes others.	High degree of realism and replication possible.	Limited to population and community-level studies.	Grazing and predation experiments, experiments on rocky shores.
Field mesocosms	1–$100\ m^3$	Walls prevent exchange of water, organisms, and materials between what is inside and outside of container.	Intermediate degree of control and realism. Easy to maintain. Less costly than land-based systems. High degree of replication possible.	Difficult to simulate realistic internal mixing. Isolation and artifacts of scale can distort dynamics.	Limnocorrals Controlled Ecosystem Population Experiment (CEPEX).
Laboratory mesocosms	0.01–$100\ m^3$	Walls isolate organisms and processes. Chemical and physical environment is maintained through technology.	Very high degree of control over internal biological and physical conditions. Convenient location allows continuous monitoring and manipulation. High degree of replication possible.	May be expensive to construct and maintain depending on the degree of environmental control. Isolation and artifacts may distort dynamics.	Multiscale Experimental Ecosystem Research Center (MEERC); Marine Ecosystem Research Laboratory, University of Rhode Island (MERL); Ecotron Controlled Environmental Facility.
Physiological-scale laboratory experiments	0.001–$1\ m^3$	No attempt made to simulate an ecosystem.	Allows highly controlled study of individual processes.	Extrapolation to ecosystem context can be difficult.	Photosynthesis vs. irradiance experiments sediment flux rate experiments.
Simulation models	Matched to research question and data	Create mathematical analog of ecological system. Ask "what-if" questions.	Infinite control over what is included. Useful tool for both designing experiments and extrapolating results.	Realism of output is entirely constrained by assumptions: garbage in = garbage out.	Ecopath Ecosim

Enclosed experimental ecosystems have a unique role in coastal marine research

Mesocosms are widely used to conduct experiments on a broad range of terrestrial and aquatic habitats (Fig. 10). Their use has become particularly important in estuarine ecosystems because of the unique challenges and limitations associated with manipulative field research in these complex systems.

These challenges and limitations occur because estuaries are open ecosystems with high degrees of biological, material, and energetic exchange with both the land and the sea. Although there is a net flow of water from land to sea, flood tides bring organisms and water from the sea upstream into the estuary. In this way, the tidal cycle creates strong gradients in the abundance of organisms and in biogeochemical processes from the mouth of the estuary to its upper reaches. Species of organisms that are relatively fixed in location during all or part of their lifecycle, such as

aquatic grasses and some bottom-dwelling invertebrates, experience constantly changing conditions with the ebb and flow of tides and shifting runoff from land. In contrast, planktonic organisms travel passively with the water and are strongly affected by physical processes such as vertical and horizontal mixing. Finally, mobile organisms, such as fish and certain crustaceans, are able to selectively choose favorable environmental conditions within the estuary. All of these organisms are influenced by wind patterns, storms, and daily and seasonal cycles of materials that generate continually changing conditions. These same conditions render controlled field-based experimentation challenging. Given these dynamic attributes of estuaries, certain types of experiments are only feasible in the more controlled environmental conditions provided by enclosed experimental ecosystems.

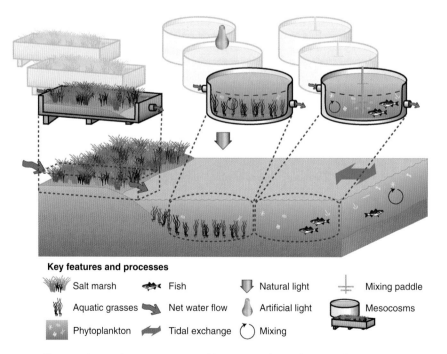

Key features and processes

Salt marsh	Fish	Natural light	Mixing paddle
Aquatic grasses	Net water flow	Artificial light	Mesocosms
Phytoplankton	Tidal exchange	Mixing	

Figure 10: In nature, all types of coastal ecosystems are subject to complex exchange with surrounding ecosystems, making controlled experimentation in the field challenging. Mesocosms are used to conduct more controlled experiments in a variety of coastal habitats including salt marshes (left), aquatic grasses (middle), and pelagic-benthic systems (right).

Enclosed experimental ecosystems can be used to advance theory, support management, and enhance education

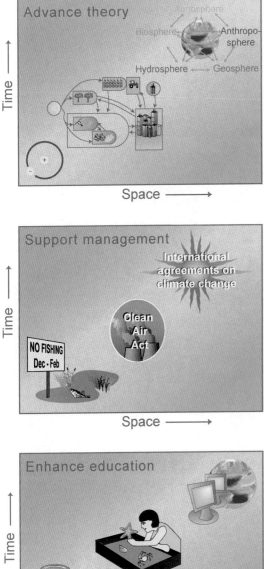

Figure 11: Advancing ecological theory, managing natural resources, and educating students all involve issues of scale and all can benefit from use of enclosed experimental ecosystems.

Ecologists use the results of mesocosm studies to assess the validity of their hypotheses and to make models of the way the world works more realistic. In turn, the results of models suggest new hypotheses that can be tested using mesocosm experiments. Iterative back-and-forth comparisons between theoretical models, mesocosm research, and field-based research are a powerful approach to advancing our understanding of environmental science (Fig. 11, top panel).

Ecosystem-based management of coastal resources depends on a thorough understanding of how the various components of the resource or fishery interact with each other and respond to human activity (Fig. 11, middle panel). Field observations and small-scale laboratory studies are necessary but insufficient to develop this understanding. Experimental ecosystems can be used to get a more complete answer to questions such as: What happens to living resources if nutrient input to the coast increases or decreases? How might foreign invasive species compete with economically important native stocks? What factors are responsible for increases and decreases in the abundance of native stocks?

Mesocosms have also been widely and successfully used at educational levels from grammar school through graduate training as a hands-on tool for teaching ecology (Fig. 11, lower panel). Simple aquatic ecosystems in mason jars can help youngsters understand the fundamental interdependencies among plants, animals, and physical factors. More complex models of marshes and other specific habitats can help students explore how different ecosystems perform different kinds of ecological services and have different sensitivities to human disturbance. When appropriately employed, mesocosms stimulate students to ask and to answer "what-if" questions. Thus, they are fundamental tools for developing scientific literacy.

Education and research can be combined in experimental ecosystem research

Public exhibits of coastal organisms are composed of living and nonliving components that are designed, constructed, and operated for education and entertainment. Such exhibits are important because they provide direct access to organisms and ecosystems that the general public would not otherwise experience. Because the emphasis is on visual accessibility, coastal exhibits are typically more like zoos and museums than functional ecosystems.

One approach to making these exhibits more dynamic and instructive models of nature is to add additional living and nonliving components of the natural ecosystem being represented—essentially to make them enclosed experimental ecosystems. Specifically, adding habitat for microbial communities, adequate light for photosynthesis, and realistic habitat and turbulence contributes to more realistic displays with higher educational value.

Value can be maximized in these systems by combining research with education. Examples include the living coral reef systems once maintained at the Smithsonian Institution. The algal-scrubber technology that was developed for this system to remove nutrients and maintain healthy living coral contributed to scientific understanding of the important interactions between coral reefs and adjacent ecosystems in nature. Similarly, exhibits like the Reef HQ in Australia, the kelp forest exhibit at the Monterey Bay Aquarium, USA, and the Wild Coast exhibit in Canada can be used by researchers studying behavioral or reproductive ecology among the fish or invertebrate species living in the enclosed experimental ecosystem. Signs and interactive displays at these and other combined research and educational facilities (Fig. 12) enhance instructional value by explaining the ecological technologies employed and research being conducted.

Figure 12: Examples of facilities that combine research and education: (*a*) Biosphere 2, *an airtight replica of earth's environment, Tucson, Arizona;* (*b*) Reef HQ, *the largest living coral reef aquarium in the world, Townsville, Australia;* (*c*) *Three-story* Kelp Forest *exhibit, Monterey Bay Aquarium, California;* (*d*) Wild Coast *exhibit, Vancouver Aquarium Marine Science Centre, Vancouver, Canada.*

Concepts of scale in ecology

Extrapolating findings from enclosed experimental ecosystems to nature is challenging

Ecological theory cautions that certain patterns and processes change as scale is increased, so it may not be possible to directly apply findings from experimental systems to larger-scale natural systems. For instance, one might add nutrients to a small mesocosm to assess the effects of agricultural runoff on plankton community dynamics. It is possible that the bloom of phytoplankton in this mesocosm is a close match to the response when similar concentrations of nutrients are added to a natural coastal ecosystem. However, because the experimental ecosystem is smaller, shallower, and less ecologically complex than nature, it is also possible that the bloom in the mesocosm will differ in timing, magnitude, and species composition from what would occur in nature.

These potential differences in response between mesocosms and nature do not invalidate research in enclosed experimental ecosystems, but instead challenge researchers to develop quantitative and systematic approaches to extrapolate findings across a range of scales (Fig. 13). Researchers need to use methods for designing scale-sensitive experiments, the results of which can be systematically extrapolated to nature. The material in the following pages examines the meaning of scale and presents what scientists have learned about scale and its significance in ecology.

Figure 13: Pelagic-benthic habitats (upper), aquatic grass habitats (middle), and salt marsh habitats (lower) can all be simulated in mesocosms. However, there is not necessarily a one-to-one correspondence between the dynamics observed in mesocosms and nature; researchers and managers need to understand how to extrapolate information from experiments to nature.

The term scale can be used as a noun to define critical experimental characteristics

Discussions of ecological scale can be confusing because *scale* means different things to different researchers and in different contexts; a meaningful discussion requires a clear definition (Petersen et al. 1999). Among other things, the word scale is two parts of speech, a noun and a verb. As a noun, *scale* refers to a set of variables that can be used to characterize time, space, and complexity of organisms, ecosystems, and experiments (Fig. 14; Petersen et al. 1999). For instance, two key aspects of temporal scale that must be considered in the design of experiments are duration (e.g., length of experiment, organism generation time, perturbation response time) and frequency (e.g., timing of perturbations, sampling interval, exchange rates, variability in the timing of inputs; Table 2). Spatial scale is defined by variables such as length, depth, area, volume, shape, and heterogeneity (patchiness).

Complexity lacks the clearly definable dimensions associated with temporal and spatial scale. Nevertheless, the concept of complexity scale can be applied to characterize such variables as species diversity, feeding relationships (e.g., trophic depth vs trophic breadth), physical diversity of the environment (number of habitats, degree of connection between habitats, diversity of biogeochemical environments, etc.), and levels of ecological organization under consideration (population, community, or ecosystem; Fig. 15). A first step toward addressing scale in ecology is for researchers to clearly define and understand the scales of time, space, and complexity (Table 2).

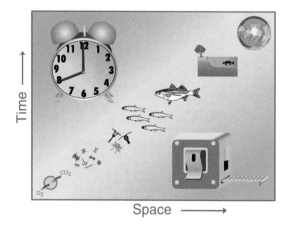

Figure 14: Ecological scales include time and space.

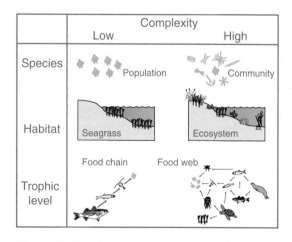

Figure 15: Complexity is an aspect of ecological scale.

Table 2: Different variables need to be considered for different components of scale.

Scale	Variables that must be considered in experimental design
Time	Experimental duration, organism lifespan, sampling frequency, and frequency of manipulation
Space	Length, depth, area, volume, shape, and patchiness
Time–space interactions	Speed (length/time), and exchange rate (length^3/time)
Ecological complexity	Species or functional group diversity, levels of organization, food webs, and habitat diversity

The term scale can be used as a verb to describe the process of extrapolating information

The verb *to scale* refers to the act of relating or extrapolating information from one space, time, or complexity to another. Quantitative measures are necessary to scale or translate information from small experimental systems and observations to large natural systems, from short duration experiments and measurements to longer time periods, from parts of nature contained in experiments to whole ecosystems, and from simplified experimental ecosystems to the full complexity of nature (Fig.16). Understanding how to relate observations among scales is essential to maximizing the value of ecological research. Scaling patterns can be elucidated by observing such patterns in nature, by developing ecological theory, and

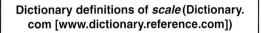

Dictionary definitions of *scale*(Dictionary. com [www.dictionary.reference.com])

1. Scale (noun). Relative dimensions without difference in proportion of parts; a distinctive relative size, extent, or degree; the ratio between the size of something and a representation of it

2. Scale (verb). To pattern, make, regulate, set, measure, extrapolate, or estimate according to some proportion, rate, standard, or control

Scale – from Latin *scala* for ladder, stairs

by conducting experiments that are explicitly designed to assess the effects of scale.

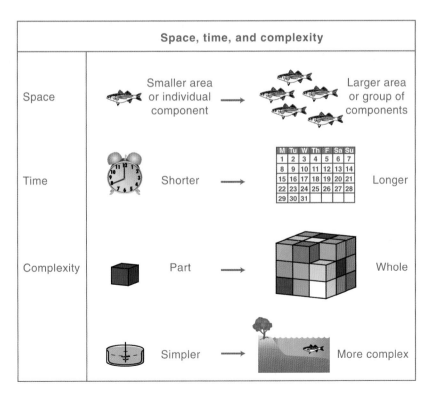

Figure 16: *Aspects of space, time, and complexity can scale from a minimal attribute to maximum or compound attributes.*

Ecological scale is characterized by both grain and extent

Landscape ecologists have emphasized two additional aspects of scale, namely *grain* and *extent* (Turner and Gardner 1991; Fig. 17). Grain refers to the smallest size or time scale that can be resolved. By analogy, it is useful to think of grain in terms of the pixels per square inch of film (spatial grain) and the frames per second shot by a movie camera (temporal grain). On the other hand, extent is the total size and time scale of interest, and is analogous to the size of the scene captured on the film (spatial extent) and to the total length of the movie (temporal extent).

In general, there are three contexts for scale in ecological research (Kemp et al. 2001; Table 3). Observational scale (Fig. 18, top panel) describes the scales that a scientist is able to see through the particular research tool that is employed. Grain is the smallest scale that is detected with the technology, and extent is the largest scale that is detected. New technologies allow us to detect ever smaller and ever larger scales. Observational grain and extent are solely determined by the data collection method and do not necessarily reveal anything about underlying structure and ecological processes. For instance, the grain and extent of satellite imagery are determined by the camera technology (grain is the resolution or pixel size and extent is the area coved by the entire image captured). Most landscape ecology literature uses

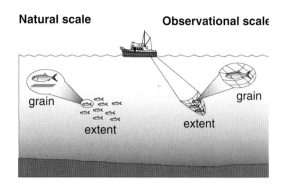

Figure 17: *The length of a certain species of fish and the average size of a school of that species are examples of natural or characteristic grain and extent respectively. The mesh size of a fishing net represents the grain of observational scale used in a fish capture study. The width of the entire net, the distance towed, and the duration of the tow are all components of observational extent.*

grain and extent in this context. Theory suggests that great care must be taken when making inferences about dynamics that operate at scales finer than observational grain size or broader than observational extent (Wiens 1989).

Experimental scale (Fig. 18, middle panel; Table 3) describes the scale of an experimental system: Grain is sample size, sampling interval, and frequency. Extent is measured by the size, duration, and complexity of an experimental system. As with observational scale, there are constraints when making inferences below

Table 3: *Three contexts for scale in ecological research.*

Scale	Grain	Extent
Observational	Instrument resolution	Total area and duration observed
Experimental	Sample size and frequency	Experimental unit size (e.g., mesocosm volume, width, and depth) and experimental duration
Natural or Characteristic	The size and time scale of organisms and unit processes (e.g., size and lifespan of an individual organism)	The size and time scales of groups of organisms and complete processes (e.g., a school of fish), size, and turnover time of an estuary

Ecological scale is characterized by both grain and extent

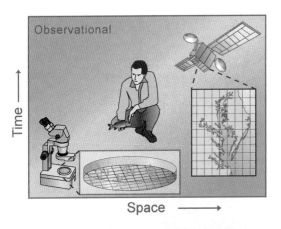

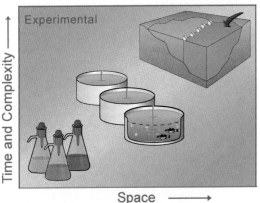

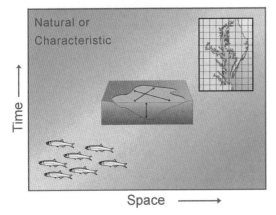

Figure 18: There are three distinct ecological contexts for the terms grain and extent (Wiens 1989) that vary depending on whether the data are measured in nature (observational scale), collected through experimental manipulations (experimental scale), or detected as intrinsic characteristics of a natural system (natural or characteristic scale).

experimental grain size. However, unlike observational scale, experimental studies almost always seek to extrapolate inferences beyond the scale of experimental extent. For instance, one might use the results of a 30-day experiment conducted in 100 L containers to draw inferences about phenomena that occur within a coastal bay over an entire season. The need to extrapolate beyond experimental extent can be viewed as the crux of the scaling problem faced by experimentalists.

Natural or characteristic scale (Fig.18, lower panel; Table 3) describes natural scaling boundaries associated with organisms, processes, and ecosystems. Organism size, home range, lifespan, basin depth and volume, and watershed area are all natural scales. For example, in a school of fish, natural grain would be defined as the size and generation time of an individual fish whereas natural extent would be the size and longevity of the school itself. Natural scale differs from observational and experimental scales in that dimensions are defined by objectively identifiable boundaries that exist in nature (Fig. 18). Ecosystem components with similar natural grain or extent are likely to interact more strongly with each other than with components with different natural scales.

Natural scales can be identified by systematically varying observational grain and extent (Wiens 1989); rapid changes and discontinuities in the measured pattern or process that occur over small changes in observational or experimental grain and extent indicate the presence of boundaries associated with natural scales (Krummel et al. 1987).

Scale is dependent upon the perspective of the researcher

In observational, experimental, and natural scales, grain and extent are dependent on the phenomena under investigation or on the technology available for investigation. The definitions of grain and extent and the scale used to define grain and extent change according to the needs of the researcher and the particular question being addressed. For example, for a researcher focused on population dynamics within a given salt marsh, the stand of salt marsh plants within that marsh represents experimental extent. In contrast, for a researcher focused on understanding the role of salt marshes in the estuary as a whole that same area represents experimental grain. For yet another researcher examining the coastal zone, the salt marsh is the grain within a much larger experimental extent.

From the perspective of a population ecologist working with sessile organisms, a small (e.g., 10 m x 10 m) plot and a specific season (e.g., summer) may define the experimental extent scale (Fig. 19, top panel). In contrast, from the perspective of an ecosystem scientist, the same plot and time may represent experimental grain (e.g., the scientist samples many 10 m x 10 m plots once per season), and the size of the entire watershed and several years might appropriately define both experimental and characteristic extent (Fig. 19, middle panel). From the perspective of a global change scientist, that same watershed may represent experimental and characteristic grain size (i.e., single pixel) in a model that defines extent as the regional landscape or even the whole biosphere over decades or centuries (Fig. 19, bottom panel).

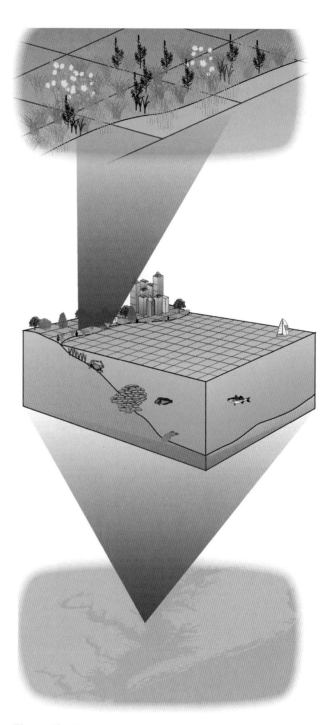

Figure 19: *Three representative scales arranged from smallest (top) to largest (bottom).*

Mesocosms and scaling patterns

There is a long history of the use of experimental ecosystems

The initial concept of microcosms, as hierarchically nested miniature worlds contained successively within larger worlds, has been credited to early Greek philosophers including Aristotle (Beyers and Odum 1993). Although it is difficult to date the initial scientific uses of enclosed experimental ecosystems, their use in terrestrial ecology occurred in the early 1900s. Small glass jars and other containers were widely used as experimental ecosystems by the middle of the twentieth century (Beyers and Odum 1993; Edmondson and Edmondson 1947; Odum 1951; Whittaker 1961). H. T. Odum and his colleagues were pioneers and proponents of the use of mesocosms to study aquatic ecosystems. They constructed a wide variety of experimental ecosystems including laboratory streams (Fig. 20) (Odum and Hoskin 1957), containers with planktonic and vascular plant communities (Beyers 1963), and shallow outdoor ponds containing oysters, seagrasses, or both (Odum et al. 1963). Although the word *microcosm* was used initially to describe virtually all experimental ecosystems, *mesocosm* was later adopted to distinguish larger experimental units from smaller bench-top laboratory systems such as chemostat microcosms (Fig. 21) (Odum 1984).

Some have suggested that experimental manipulations of whole aquatic ecosystems

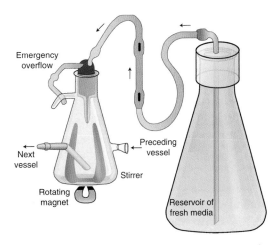

Figure 21: Chemostat microcosms have been widely used to measure growth rates of phytoplankton (Margalef 1967).

in nature are always preferable to mesocosm studies (Schindler 1998). However, the large spatial scale, two-way tidal water exchanges, and natural variability make such whole-ecosystem manipulations extremely difficult in natural coastal ecosystems, leaving mesocosms as a critical tool for controlled experimentation (Kemp et al. 2001). A series of books devoted to aspects of experimental aquatic ecosystems mark recent progress with this research approach (Beyers and Odum 1993; Giesy 1980; Grice and Reeve 1982; Lalli 1990; Graney et al. 1994; Gardner et al. 2001).

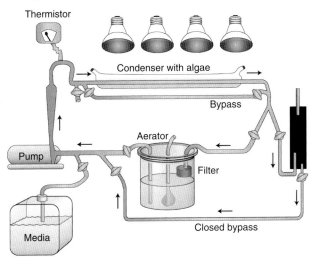

Figure 20: Flowing stream microcosm similar to those used by Odum. The system was normally open to gas exchange but closed during metabolic measurements. Temperature and stream velocity were controlled (Odum and Hoskin 1957).

There are diverse styles and applications of experimental ecosystems

In the decades since the pioneering research of Odum and his associates, experimental microcosms and mesocosms have been developed in a diversity of sizes, shapes, and habitats to address a broad range of research questions (Petersen et al. 1999). For example, small 0.5 L chemostat flasks have been widely used to study plankton in the laboratory (Margalef 1967) (Fig. 21), while large plastic bag enclosures have been employed to study communities *in situ* in lakes and the ocean (Grice et al. 1977). Similarly, mesocosm shapes vary from tall and narrow (16 m high×2.5 m diameter) plankton towers (Menzel and Steele 1978) to broad (350 m² surface), shallow (1 m deep) estuarine ponds (Twilley et al. 1985). Aquatic mesocosms have been constructed to study diverse habitats, including planktonic regions of lakes (Kemmerer 1968), oceans (Menzel and Steele 1978), estuaries (Nixon et al. 1984), deep benthos (Kelly and Nixon 1984), shallow ponds (Goodyear et al. 1972), streams (McIntire 1968), coral reefs (Williams and Adey 1983), salt marshes (Fig. 22) (Kitchens 1979), and seagrasses (Twilley et al. 1985). Composition and organization of experimental ecological communities range from simple *gnotobiotic* ecosystems where all species are selected and identified (Taub 1969; Nixon 1969), to interconnected microcosms containing different trophic levels (Fig. 23) (Ringelberg and Kersting 1978), to intact undisturbed columns of sediment and overlying water (Kelly and

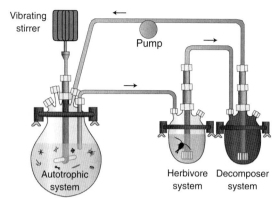

Figure 23: Three-trophic-level microcosm for study of feeding relationships and general patterns of ecosystem development (Ringelberg and Kersting 1978).

Nixon 1984), to tidal ponds with self-organizing communities developed by seeding with diverse inoculant communities taken from different natural ecosystems (Odum 1989).

Mesocosms have been used effectively to address theoretical questions, succession (Cooke 1967), homeostatic control of ecosystem production (Copeland 1965), species diversity and community stability (Reed 1978), feedback effects of herbivore grazing (Cooper 1973), and top-down versus bottom-up control (Threlkeld 1994). As tools for assessing environmental effects, mesocosms have also been widely used to test for ecosystem responses to nutrient enrichment (Kemmerer 1968; Nixon et al. 1984; Oviatt et al. 1979) and to inputs of toxic contaminants (Cairns 1979; Graney et al. 1994; Taub 1997).

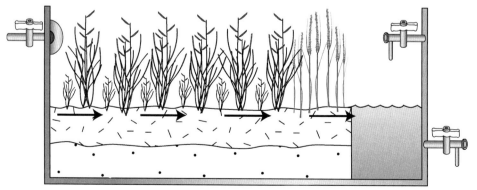

Figure 22: Marsh mesocosms have been used to explore a range of issues including species diversity and community stability (Kitchens 1979).

Scale poses challenges for experimental ecosystem research

It has long been recognized that scale is an inherent problem that tends to confound interpretation of results from experimental ecosystem studies (Kemp et al. 1980; Pilson and Nixon 1980; Gamble and Davies 1982). Small enclosures and short experiments limit the number of trophic levels included in an experiment (Gamble and Davies 1982; Grice 1984). Since mesocosm use first became prevalent in the 1970s (Fig. 24), researchers have expressed concern that artifacts associated with enclosure (e.g, wall growth, animal movements, mixing regimes, and water exchange rates) may be related to the small scale of experimental ecosystems (Harte et al. 1980; Kuiper 1982; Bloesch et al. 1988). Nevertheless, the majority of mesocosm studies have ignored the question of scaling and the problem of extrapolation (Petersen et al. 1999). Although a few investigators have used a simple idea of mesocosm calibration by which key properties are adjusted in experimental systems to mimic conditions in the natural environment (Nixon et al. 1980), no widely-accepted quantitative scaling or extrapolation method has existed to aid in design and interpretation of mesocosm studies.

By the end of the 1980s it was clear that further progress in the application of experimental ecosystem methods to aquatic science would require focused quantitative study of how scale affects behavior in natural and experimental ecosystems and how results from experimental ecosystem studies might be systematically extrapolated to conditions in nature at large (Grice 1984; Kimball and Levin 1985; Frost et al. 1988; Petersen and Hastings 2001). A first step in the process of addressing the important issues of scale was to assess the scaling attributes of prior research conducted in aquatic mesocosms. The next several pages discuss the findings of a comprehensive review of prior experiments.

Figure 24: (*a*) *Loch Ewe enclosed ecosystems program, Scotland.* (*b*) *Rocky littoral mesocosms at Solbergstrand, Norway.* (*c*) *Marine Ecosystem Research Laboratory, University of Rhode Island.*

Scaling issues in the design of enclosed experimental ecosystems are often overlooked

The literature shows a lack of attention to issues of scale in experimental ecology. A quantitative review of the scaling attributes of 360 studies conducted in aquatic mesocosms found that only 43% of published papers provided complete physical dimensions (width, depth, volume,

Vic Kennedy

Sandy Rodgers

Figure 25: *A row of 1 m³ indoor pelagic-benthic tanks, Multiscale Experimental Ecosystem Research Center* (left)*; An outdoor mesocosm in the Patuxent River, Maryland* (right)*.

duration) of the experimental systems (Table 4) (Petersen et al. 1999). This lack of information about experimental scale makes it exceedingly challenging to apply research findings to nature.

Several scaling problems can be identified. The most obvious problem is reductions in the scale relative to nature. The median duration of 49 days and median volume of 1.7 m³ found in mesocosm studies (Petersen et al. 1999) are brief and small relative to the scales that characterize many ecological processes in coastal ecosystems. For reasons that stem in part from these reductions in physical scale, components of ecological complexity such as trophic depth (food-chain length) and breadth (number of species at a trophic level) are also typically reduced in experimental systems relative to the larger natural ecosystems that they are intended to represent (Fig. 25).

Table 4: *Statistics that describe scale related attributes of research conducted using experimental ecosystems. The* Experiments reporting *column shows the fraction of the 360 published papers examined that reported a value for a variable. The* Complete dimensions *row describes the percentage of papers that supplied a complete set of values for either length, width, or depth (rectangular mesocosms) or radius and depth (cylindrical mesocosms). Acclimation is the pre-treatment acclimation period. The 25th and 75th percentiles* (Interquartile range) *are provided as a measure of spread (Petersen et al. 1999). n/a= not available*

Variable	Median values	Interquartile range 25%	75%	Experiments reporting (%)
Length (m)	1.2	0.3	8.0	16
Width (m)	0.9	0.3	2.3	20
Radius (m)	0.9	0.5	1.5	34
Depth (m)	2.4	0.5	5.5	61
Complete dimensions	n/a	n/a	n/a	43
Volume (m³)	1.7	0.03	15.2	84
Duration of experiment (d)	49	20	120	74
Number of treatments	2	1	3	57
Number of replicates per treatment	2	1	3	65
Number of controls	2	1	3	59
Acclimation (d)	27.5	7.0	60.0	19
Volume sampled (L)	1.0	0.4	25.0	6

Aquatic mesocosms are typically operated for weeks to months and involve a wide range of volumes

A review of the literature reveals clear patterns in the size and duration of enclosed experimental ecosystems (Petersen et al. 1999) (Fig. 26). Although a peak in frequency is evident for 1-year studies, overall the data roughly approximate a log-normal distribution (Fig. 27, upper panel). A number of studies (5%) last 1 year or longer, but 50% of the experiments examined lasted between 20 and 120 days and more than half were conducted for periods of 50 days or less.

These data indicate that the effects of ecological processes operating on time scales longer than one season or one year are not typically captured in mesocosm studies. This finding is of concern because responses observed in short-duration experiments (field as well as mesocosm) may not adequately characterize longer-term trends. Indeed, evidence has emerged that in many cases short-term dynamics may be quite distinct from, or even opposite to, the long-term dynamics that occur in response to a given treatment (Frost et al. 1988; Tilman 1989).

Although experimental duration of studies ranges from days to thousands of days, container volume ranges over a factor of 10^{16}, from 8×10^{-12} m^3 microcapillary tubes used to study dynamics in a ciliate community (Have 1990) to 1.8×10^4 m^3 *in situ* experimental tubes used to explore limnetic community interactions (Smyly 1976). As with duration, volume roughly approximates a log-normal distribution (Fig. 27, lower panel). An anomalous abundance of papers in the 10–100 m^3 range in container volume is in part a result of the prolific record of experimentation and publication associated with over 20 years of continuous research at the MERL mesocosm facility at the University of Rhode Island (Fig. 24) (Nixon et al. 1984; Oviatt et al. 1979).

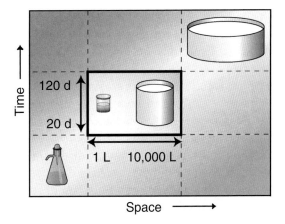

Figure 26: The frequency distribution of scientific papers on mesocosms indicates a wide range of volumes used, but a smaller range of time scales.

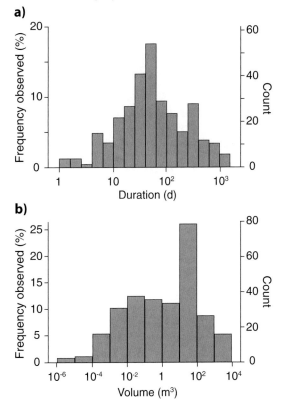

Figure 27: Frequency distribution of scientific papers describing the (a) experimental duration and (b) volume of enclosed experimental ecosystems (Petersen et al. 1999).

Selection of experimental duration does not appear to be systematically related to habitat or treatment

Since scales of time, space, and complexity are related in natural ecosystems (Fig. 28), one might expect that scaling variables would also be linked in the design characteristics of experimental ecosystems. For example, researchers might set experimental duration in relation to the characteristic scales of the organisms and processes under investigation, such as the average life time and reproductive cycles of dominant organisms, and the turnover times of energy and nutrients (Frost et al. 1988). Based on the size and associated time scales of the organisms that dominate different habitats, one might expect experimental duration to increase from plankton to benthic to pelagic-benthic to aquatic grass to marsh habitats. In contrast to this expectation, the literature reviewed reveals that median durations for marsh and plankton experiments are not significantly different from each other, but are lower than those for other habitats (Fig. 29a).

The short durations of experiments in marsh and aquatic grass habitats relative to the pelagic-benthic habitat (Fig. 29a) are logically counterintuitive given that the former are typically dominated by vascular plants that have long generation times

(months-years) relative to the generation times and characteristic bloom durations of planktonic organisms (1–100 days). At a median duration of 30 days, most marsh experiments reported in the literature are only capable of exploring immediate treatment effects rather than chronic effects that occur on time scales associated with complete producer life cycles.

Small differences in duration were also evident among treatment types (Fig. 29b). The significantly longer duration of nutrient enrichment experiments may reflect an interest in the chronic effect of eutrophication such as shifts in species composition and habitat quality (Taylor et al. 1995; Verdonschot 1996). On the other hand, the relatively short duration of toxic contaminant experiments suggests more of an emphasis on acute effects such as mortality and growth of individual organisms (Maund et al. 1992; Taylor et al. 1994). Because release of persistent contaminants into the environment is in many cases a chronic phenomenon (Davis 1993; Weber 1994), there seems no logical justification for the relatively short duration of experiments dealing with toxins.

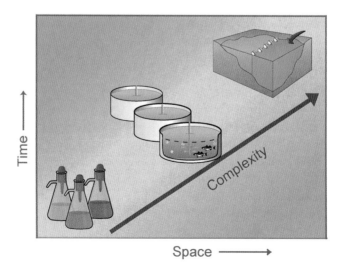

Figure 28: One might expect a clear and consistent relationship between time, space, and complexity scales in the design of experimental systems. This is not observed in a review of the literature (Petersen et al. 1999).

Selection of experimental duration does not appear to be systematically related to ecological complexity

Scaling theory suggests that the temporal and spatial scales necessary to observe ecological dynamics increase with increasing ecological complexity. Larger experimental systems and longer duration are necessary with increases in the level of ecological organization under investigation (i.e., organism < population < community < ecosystem) and with increasing number of trophic levels included (O'Neill et al. 1986). The literature review revealed that the median experimental duration did in fact increase from organism to population to ecosystem level studies, with statistically significant differences between organism and population levels (Fig. 29c). However, community level studies were inconsistent with this pattern, and this probably stems from the fact that many community level studies are conducted with small organisms (often protists) in small mesocosms.

Although differences were not statistically significant, the slight trend of increasing median duration with increasing number of trophic levels under investigation observed (Fig. 29d) is consistent with intuitive and theoretical expectations (O'Neill 1989; Beyers and Odum 1993).

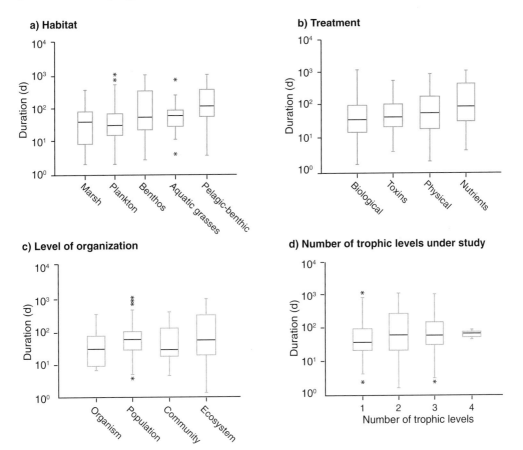

Figure 29: *Box plots of duration of mesocosm experiments and (**a**) habitat, (**b**) experimental treatment, (**c**) organizational level of response variable, and (**d**) number of trophic levels. Data are a sampling of 266 published papers describing research conducted in enclosed experimental ecosystems. Median values are represented by the bar within a box, and the 75th and 25th percentiles (i.e., the interquartile range) by the top and bottom of the box. The ends of the whiskers represent the farthest data point within a span that extends 1.5 × the interquartile range from the 75th and 25th percentiles. Data outside this span are graphed with asterisks (Petersen et al. 1999).*

Selection of experimental volume does not appear to be systematically related to other design decisions

As with temporal scale, one might anticipate a relationship between mesocosm size and characteristic ecological spatial scales (typical organism size, patch size, resource distribution). If mesocosm size is based on organism size, then mesocosm size should be smaller for planktonic systems than for habitats dominated by vascular plants. However, the literature review reveals that median volumes are significantly larger for planktonic and pelagic-benthic mesocosms than for marsh, aquatic grass, or benthic mesocosms. Marsh, aquatic grass, and benthic mesocosms are not significantly different from each other (Fig. 30a). There also are differences in the volume of mesocosms for different experimental treatments (Fig. 30b). The fact that both median duration (p. 26) and median volume for nutrient enrichment treatments are significantly larger than for toxicological treatments indicates that findings

from mesocosm experiments may be relatively biased towards smaller scales in contaminant research.

Median volume increases significantly between organism and population levels as expected from scaling theory (O'Neill 1989; Beyers and Odum 1993), but median volume for community level studies is significantly lower than all others (Fig. 30c). The small median size and short duration of community experiments may be due to the prevalent use of small defined, or gnotobiotic (Taub 1969), microbial mesocosms that have been widely used as models to explore general aspects of community dynamics (Luckinbill 1974; Have 1990, Drake et al. 1996). Consistent with scaling expectations, there is a significant increase in volume between mesocosms that include only the first trophic level (producers) and mesocosms that included up to the third trophic level (predators or grazers, Fig. 30d).

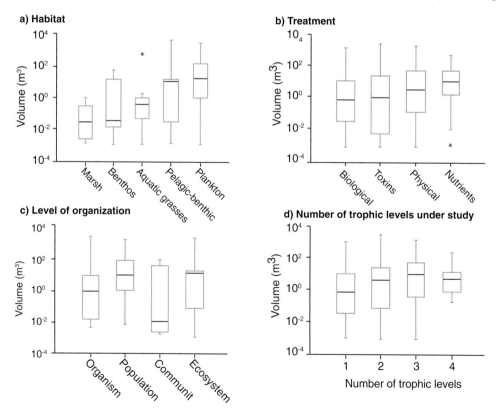

Figure 30: *Box plots of mesocosm volume and (**a**) habitat, (**b**) experimental treatment, (**c**) organization level of the response variable, and (**d**) number of trophic levels in the experiments. Data source and boxes are as in Fig. 29 (Petersen 2001).*

There is a trade-off between experimental ecosystem volume and number of replicates

Statistical power, the ability to detect significant differences among treatments, increases with increasing numbers of replicated experimental units. The literature review indicates a significant negative relationship between the volume of experimental ecosystems and number of replicates employed in experiments (Fig. 31a). This is consistent with findings in other types of ecological experiments (Kareiva and Andersen 1988). The statistical implications of this pattern are potentially important. If random variability were similar in small and large volume mesocosms, then the decrease in replication with increasing volume would be associated with a concomitant decrease in statistical power. This would then bias our understanding, leading us to understand that small-scale processes are more important than larger-scale processes. There is some anecdotal evidence that stability and similarity between replicates may increase with increasing mesocosm volume (Giddings and Eddlemon 1977; Ringelberg and Kersting 1978), implying statistical justification for decreasing

replication with increasing size, but the generality and statistical implications of the few reports remain to be determined. Nevertheless, a decrease in variability with increasing size will never allow for the application of traditional statistics to the many experiments conducted with no replication.

The number of experimental treatments also exhibits a general (though nonsignificant) trend of decrease with increasing volume (Fig. 31b), indicating that experiments conducted in large experimental ecosystems generally do not examine interactive effects. Again, the major reasons for these tendencies toward decreasing replication and number of treatments with increasing volume are probably constraints associated with feasibility rather than fundamental ecological considerations of scale. Operating a few dozen 30-L aquaria is easier than operating a single 100 m^3 enclosure. In contrast to volume, experimental duration does not vary significantly with the number of replicates per treatment (Fig. 31c) or with the number of treatments applied (Fig. 31d).

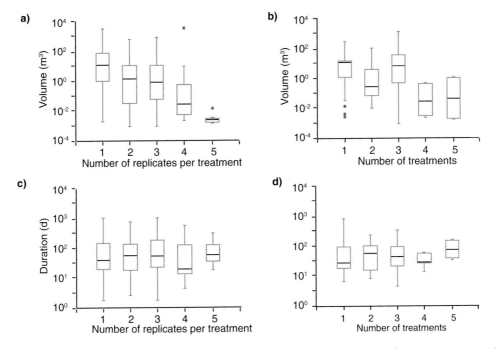

Figure 31: *Box plots of container volume and experimental duration versus number of replicates per treatment (a and c respectively) and number of treatments (b and d respectively). Boxes are as in Fig. 28 (Petersen et al. 1999).*

A wide variety of design decisions vary together with changes in the size of experimental ecosystems

The full spectrum of decisions regarding how aquatic mesocosms are initiated and maintained, such as control over temperature and light, the use of mechanical mixing devices, and water exchange are strongly linked to the choice of mesocosm size. Design characteristics that exhibit a significant decrease in prevalence with increasing volume include temperature regulation, regulation of intensity and duration of light, and specific selection of community composition (Fig. 32). This finding makes intuitive sense as these design characteristics are increasingly difficult to regulate as size is increased.

Design characteristics that increase in prevalence with increasing volume include wall cleaning and the use of *in situ* enclosures

(i.e., deployed within an aquatic environment as opposed to laboratory based tanks and artificial ponds). The increase in wall cleaning with increasing volume is somewhat illogical given the fact that the wall area per unit volume, and hence the relative influence of wall growth, decreases with increasing tank radius (Chen et al. 1997). Inclusion of water exchange and mechanical mixing are unrelated to size of experimental systems.

The finding that many important design decision choices vary together with selection of mesocosm size is of concern because co-variation confounds our ability to distinguish differences in experimental outcome that result from scale from differences potentially attributable to these other factors.

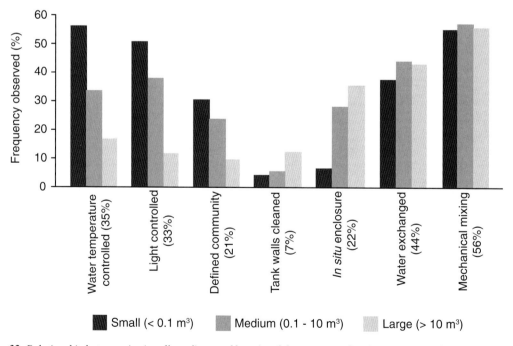

Figure 32: *Relationship between size (small, medium, and large) and the presence of various mesocosm design characteristics. Defined community indicates that individual populations were selectively added to create the mesocosm community. Water exchanged indicates the regular replacement of some fraction of water with new media or water from nature (beyond negligible amounts added to replace samples removed) (Petersen et al. 1999). The overall percentage of mesocosm experiments exhibiting a given design feature is provided in parentheses.*

Distortions in ecological dynamics result from changes in scale as a direct outcome of simple geometric relationships

The size and shape of an enclosed experimental ecosystem can have an important effect on ecological dynamics. The potential for distortions in ecological relationships as size and shape are changed is particularly evident in pelagic-benthic mesocosms. Although activities in the water column and benthic (bottom) regions are the focus of scientific research, two other regions of biogeochemical activity, the water surface and the walls, are ecologically important as well (Fig. 33). The water surface is important as the site of energy and gas exchange between the experimental ecosystem and the outside world. The walls of the mesocosm are artifacts of enclosure that are important because of the potential for the attached community of organisms to dominate biogeochemical activity.

Geometry dictates that the ratio of wall surface area to interior tank volume in a cylindrical mesocosm is inversely proportional

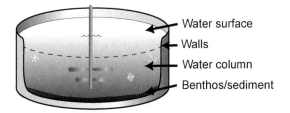

Figure 33: There are four important regions of biogeochemical activity within mesocosms.

to tank radius whereas the ratio of horizontal surface area to volume is inversely proportional to tank depth (Fig. 34). These relationships are important to experimental design because they determine the relative magnitude of the four important regions of biogeochemical activity within the mesocosms. For instance, tank radius determines the degree to which the undesirable community of periphyton that grows on the walls will influence ecological dynamics.

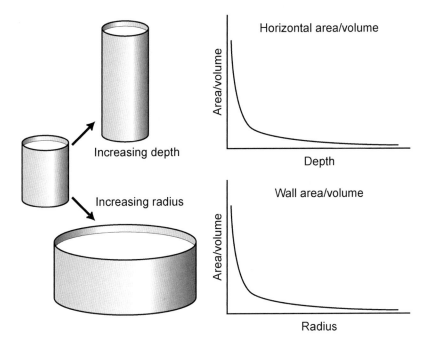

Figure 34: Geometric relations are affected by container depth and radius.

There has been a bias toward maintaining mesocosm shape in aquatic experiments

There are three alternative and mutually exclusive scaling relationships that can be held constant as container volume is increased: depth, radius, or shape (ratio of depth to radius) can be conserved (Fig. 35). An analysis of aquatic studies conducted in planktonic-benthic mesocosms reveals a statistically significant bias toward mesocosms having a common shape (Fig. 35); ecologists appear to gravitate toward a depth/radius ratio of ~4.5 (Petersen et al. 1999). There are broad consequences of this bias. In general, MEERC researchers can expect that large mesocosms are simultaneously less influenced by wall artifacts, have less benthic area per unit volume, and less surface area available for gas and light exchange per unit volume than smaller systems. Without specific understanding of scale effects, this aggregate bias toward scaling for constant shape confounds our ability to systematically compare the results of experiments conducted in different-sized mesocosms.

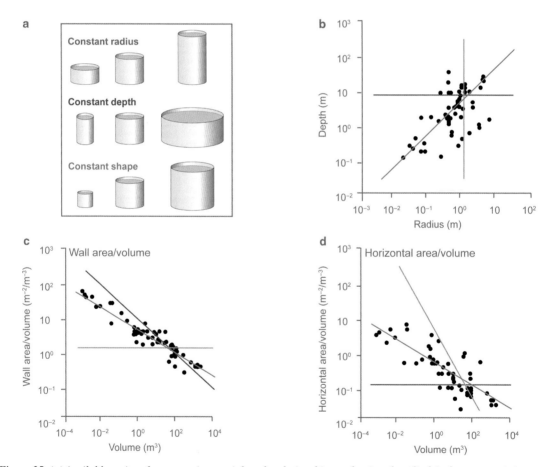

Figure 35: *(**a**) Available options for conserving spatial-scale relationships as the size of a cylindrical mesocosm is increased. (**b**) Relations between depth and radius for the cylindrical mesocosms in the ecological literature. (**c**) Surface areas of the vertical walls versus volume. (**d**) Surface area of bottom and top versus mesocosm volume. Choices represent constant depth (brown lines placed at median depth values), constant container radius (green lines placed at median radius values), and constant shape (blue lines, derived from linear regression of radius (r) on depth (z)); (Petersen et al. 1999).*

Space, time, and complexity scales tend to be linked, thus constraining scaling choices in ecosystem design

An aim of mesocosm studies is to design experimental systems that condense and distill natural relationships. As the last several pages of this book make clear, scaling distortions are inevitable. Therefore, the key question in designing experiments is, how can researchers retain key relationships in spite of these distortions?

One distortion option is to simultaneously reduce space, time, and complexity scales by extracting only the smallest organisms from a natural ecosystem for inclusion in the experimental system—for instance, including plankton but no fish (Fig. 36a). As small organisms tend to have short generation times and small home ranges (Sheldon et al. 1972), experiments conducted in small experimental ecosystems and over short time periods with these organisms are justifiable. However, for certain questions the distortion imposed by reduced trophic depth (food-chain length) in such experiments will almost certainly produce results that lack realism. For instance, this choice may exclude a variety of organisms that provide top-down control through grazing and predation.

A second option in designing mesocosm experiments is to reduce space and time scales but to maintain a desired level of trophic depth or breadth by selectively assembling complex experimental communities composed of small organisms (Fig. 36b). Many of the recent experiments designed to elucidate the effects of biodiversity on ecological function have taken this second approach (Naeem and Li 1998; Lawton 1998). A problem with this approach is that the small and sometimes early successional species used in these experiments may or may not respond in ways that represent communities composed of larger, longer-lived species that are an implicit target of experimental inference. To date, very few studies conducted in experimental ecosystem have explicitly considered how alterations in scale relative to the characteristic scales that define nature might affect ecological dynamics (Petersen et al. 1999).

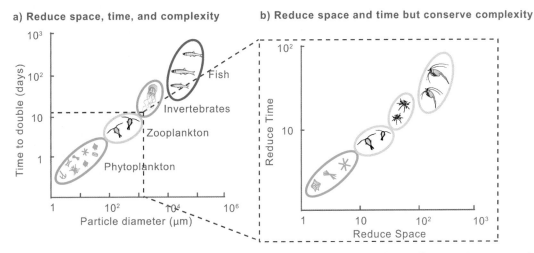

Figure 36: *(a) One option in designing experimental ecosystems is to select only the smallest organisms present in the reference natural ecosystem for inclusion (Sheldon et al. 1972). This option allows a researcher to conduct relatively rapid experiments in small experimental systems. (b) A second option is to augment the smallest organisms with other types of small organisms to maintain a greater degree of complexity. (Different circles represent different functional groups; different colors represent different size groups.)*

Scaling problems in experimental ecology are evident in field studies as well as those conducted in mesocosms

A number of critics have argued that mesocosms are too small and experiments in them are too brief and too unrealistic in complexity to adequately characterize ecological response in nature (Schindler 1987; Carpenter et al. 1995; Verduin 1969; Dewey and deNoyelles 1994; Petersen et al. 1999). These critics often suggest that experimental artifacts and scaling distortions can be avoided by using experimental manipulation of whole natural ecosystems. However, many of the scaling problems identified for mesocosms have parallels in experiments conducted in nature. For example, experimental lakes and plots (Fig.37) tend to be very small relative to natural systems on which inferences are drawn (Kareiva and Andersen 1988; Fee and Heckey 1992). Experiments conducted in large natural ecosystems are difficult to control and replicate (Fig. 38) and are often simply infeasible. Scientists are increasingly interested in understanding large-scale ecological phenomena. Yet, ecosystems, watersheds, bioregions, and the whole of the biosphere are much larger than any experiment can be (Fig. 39). Therefore, scientists need to find

Figure 37: Central and East Basins of Long Lake, Canada, are examples of experiments conducted in whole natural ecosystems. Experiments in larger natural ecosystems are challenging to conduct and to replicate.

ways to make their findings extrapolatable and generalizable beyond the scales of time, space, and complexity at which experiments are conducted (i.e., beyond experimental extent). New knowledge obtained through empirical studies is necessary to move research from qualitative to quantitative understandings of scale and toward techniques for systematic extrapolation across scales from experiments to nature.

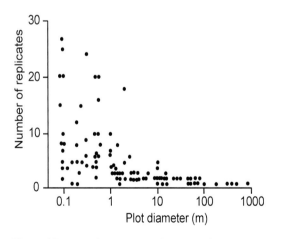

Figure 38: In studies conducted in natural ecosystems, experimental replication tends to decrease with increasing plot size. Each point is taken from a different published paper in Ecology *between January 1980 and August 1986 (Kareiva and Andersen 1988).*

Figure 39: The Ecotron *is an example of a laboratory-based terrestrial mesocosm in Silwood Park, UK. A researcher measures photosynthesis.*

Mesocosm studies and the effects of scale

Scale is fundamental to both experimentation and theory in ecological science

Experimentation and the development of new theory are fundamental and complementary approaches to clarifying the cause-and-effect relationships that underlie the workings of natural ecosystems. In the last two decades, theoreticians have significantly expanded our understanding of the importance of time, space, and complexity scales as key determinants of the ecological pattern and dynamics that are observed in nature. This trend is seen in the steady increase in the number of scientific research papers listing *scale* as a keyword (Fig. 40), and in the numerous recently published books that address scaling theory in ecology (Gardner et al. 2001; Peterson and Parker 1998).

Concurrent with advances in our understanding of scale has been an increase in the use of manipulative field and laboratory-based experiments to test ecological theory (p. 7). At this point, ecologists appear to place equal emphasis on theoretical work, field-based research, mesocosm studies, and laboratory work (Ives et al.1996).

Interest in scale and ecological experimentation are inherently linked in concept and yet remain awkwardly disconnected in practice. On the one hand, there is a heightened awareness among ecologists that the usefulness of all methods of inquiry (theory, experimentation, models) hinges on the ability to generalize conclusions drawn at particular scales and to extrapolate this new knowledge among systems that differ in temporal or spatial scales and in ecological complexity. On the other hand, experimental work in ecology has not, in general, either taken full advantage of or been effectively used to advance scaling theory.

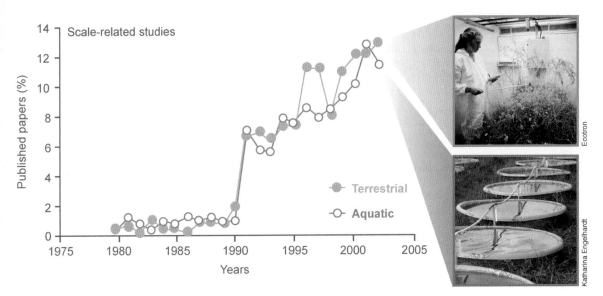

Figure 40: *Scale studies in ecology between 1980 and 2002. Separate searches were conducted by year for the term scale in keywords and abstracts of journals emphasizing terrestrial research* (Ecology, Oikos, Oecologia) *and journals publishing only aquatic research* (Limnology and Oceanography, Marine Ecology Progress Series). *The number of papers identified in each year was then standardized to the total number of papers published for that year in those journals and expressed in the graph as a percent (After Kemp et al. 2001).*

The multiscale experimental ecosystem research center (MEERC) used various habitat mesocosms to address key scaling questions

The parallel interests in scale and in manipulative experiments, coupled with concern about the relative merits of different approaches to experimentation, raise a series of questions that were the focus of long-term studies at the Multiscale Experimental Ecosystem Research Center (MEERC). These studies manipulated time, space, and ecological complexity to improve our understanding of the effects of scale on ecological dynamics. Whereas most ecological experiments are designed to quantify responses to external or internal perturbation, MEERC research was unique in that the objective was to assess scale as an independent variable driving ecological dynamics and responses to specific perturbations.

Specifically, researchers at MEERC asked the following questions: How can advances in scaling theory be used to improve the design of ecological experiments? Conversely, can multiscale experiments be designed , for example, in mesocosms of different size, shape, duration,

MEERC Research

Objectives

- Test scale-related theory
- Improve experimental design so as to simulate realistic ecosystem dynamics
- Develop rules and tools for extrapolation from experiments to nature

Approaches

- Synthesize existing information
- Perform multiscale experimentation using mesocosms and simulation models

Scope

- Coastal habitats (plankton, benthos, aquatic grasses, and marsh)
- Population, community, and ecosystem levels of organization
- Response to nutrient enrichment, trophic manipulations, and contaminants
- 10 years of US EPA funding

Table 5: *Dimensions of pelagic-benthic mesocosms in the MEERC experiments. Letters correspond with dimensions listed in figure below.*

Mesocosm	Diameter (m)	Depth (m)	Volume (m³)	Wall area (m²)
A	0.35	1.0	0.1	1.1
B	0.52	0.46	0.1	0.8
C	1.13	1.0	1	3.5
D	2.44	2.2	10	16.4
E	3.57	1.0	10	11.2

Figure 41: *Different sizes of pelagic-benthic mesocosms used in the MEERC experiments. These mesocosms were organized into three series; one with a constant depth (A,C,E), one with constant shape (B,C,D), and one with constant volume (A,B;D,E).*

or complexity to test and advance scaling theory? Can practical and empirical rules be derived to guide experimental design and to facilitate both the systematic comparison of results among experiments that differ in scale and the extrapolation of results from small-scale experiments to large natural ecosystems?

Experiments were conducted in a variety of estuarine habitats including pelagic-benthic, aquatic grass, and salt marsh systems to determine the effects of ecosystem depth, radius, size, exchange rate, experimental duration, and ecological complexity on primary productivity, biogeochemistry, and trophic dynamics.

Pelagic-benthic (PB) mesocosm experiments were used to elucidate the effects of ecosystem size and shape, material exchange rates, mixing intensity, light energy, and trophic breadth and depth. These experiments, performed over periods of days to months, used containers that ranged in volume from 0.1 m³ to 10 m³ (Fig. 41, Table 5).

MEERC used marsh, aquatic grass, and linked mesocosms to conduct experiments

Research in MEERC marsh mesocosms (Fig. 42a; Table 6) investigated the effects of plant species diversity, disturbance regimes, and differences in the quality of ground-water inputs on marsh ecosystem development. Marsh mesocosms were maintained continuously for over 7 years.

Research on experimental aquatic grass communities (Fig. 42b) examined the effect of nutrient loading, water exchange rate, and trophic complexity on competition between vascular plants and epiphytes.

Materially linked aquatic grass and pelagic-benthic mesocosms (*multicosms*, Table 6 and p. 81) were constructed to explore the effects of different exchange regimes on nutrient responses in these multihabitat ecosystems.

Environmental conditions in all types of mesocosms used in MEERC experiments were carefully controlled to simulate critical features of the estuarine environment. These experimental ecosystems nevertheless represented intentional simplification of the physical and biological complexity present in nature. The spatial extent (i.e., size) of even the largest system in these studies was smaller than patch size (i.e., smaller than characteristic grain) in nature.

Figure 42: *(a) Mesocosms with salt marsh and (b) aquatic grass.*

Table 6: *Dimensions of the salt marsh and aquatic grass mesocosms in the MEERC system.*

Mesocosm	Length (m)	Width (m)	Depth (m)	Volume (m³)
Salt marsh tanks	6.0	1.0	0.65	n/a
Aquatic grass small	0.06	0.3	0.3	0.05
Aquatic grass large	1.1	1.1	0.5	0.5
Multicosms	1.5	0.9	0.9	1.0

n/a = not applicable.

Experimental ecosystems uniquely answer questions at intermediate spatial and temporal scales

Integrated ecological research necessarily involves complementary approaches at multiple scales. Controlled laboratory experiments that focus on single species or processes are essential for elucidating ecological mechanisms. Large-scale field research is critical for testing our understanding of nature. Mesocosm research provides the crucial intermediate-scale assessment that links information gained through small-scale laboratory experiments and large-scale field research.

Reductions in scale and associated artifacts, co-variation in design attributes, and distortions in scale all pose important challenges to executing experimental ecosystem studies that accurately model conditions in nature. However, these challenges can be viewed as interesting research opportunities for advancing theoretical and practical components of the science of scale (Meentemeyer and Box 1987). Research conducted at the Multiscale Experimental Ecosystem Research Center was designed to shed light on fundamental effects of scale evident in natural (Fig. 43) and experimental (Fig. 44) ecosystems (e.g., the effects of water depth),

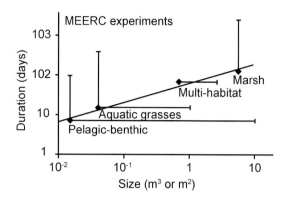

Figure 44: MEERC research was conducted in multiple estuarine habitats and spanned a broad range of time and space scales (Petersen et al. 2003).

as well as the artifacts of enclosure attributable to the artificial environment in experimental ecosystems (e.g., the effects of wall growth).

The authors of this book draw on 10 years of MEERC research to consider approaches for designing aquatic mesocosm experiments that represent nature as realistically as possible. Specific scale-related questions addressed include the following:

- What are the effects of depth and tank radius on the dynamics of experimental pelagic ecosystems?

- To what extent can effects of water depth observed in these mesocosm studies be extrapolated to nature?

- How do food-chain length and species diversity (two features of ecosystem complexity) affect estuarine ecosystem dynamics?

- How does water residence time within estuarine habitats and among different types of estuarine habitats affect ecological dynamics?

- To what extent does a system's physical scale affect variability of its ecological properties?

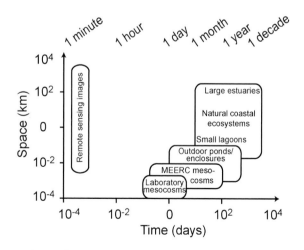

Figure 43: Approximate time and space scales relevant to coastal ecosystems that are encompassed by commonly applied research approaches, including those used in the Multiscale Experimental Ecosystem Research Center (Boynton et al. 2001).

Subsequent sections of this book are intended to describe relevant research findings and discuss the rules and tools that are available for improving the design of mesocosms experiments so that they more realistically represent the dynamics of larger-scale natural ecosystems.

References

Beyers, R.J. 1963. The metabolism of twelve aquatic laboratory microecosystems. Ecological Monographs 33:281–306.

Beyers, R.J. and H.T. Odum. 1993. Ecological Microcosms. Springer-Verlag, New York.

Bloesch, J., P. Bossard, H. Bührer, H.R. Bürgi and U. Uehlinger. 1988. Can results from limnocorral experiments be transferred to in situ conditions? Hydrobiologia 159:297–308.

Boynton, W.R., J.D. Hagy and D.L. Breitburg. 2001. Issues of scale in landmargin ecosystems. Pages 299–330 in R.H. Gardner, W.M. Kemp, V.S. Kennedy and J.E. Petersen (eds.). Scaling Relations in Experimental Ecology. Columbia University Press, New York.

Cairns, J. Jr. 1979. Hazard evaluation with microcosms. International Journal of Environmental Studies 13:95–99.

Carpenter, S.R., S.W. Chisholm, C.J. Krebs, D.W. Schindler and R.F. Wright. 1995. Ecosystem experiments. Science 269:324–327.

Chen, C.-C., J.E. Petersen and W.M. Kemp. 1997. Spatial and temporal scaling of periphyton growth on walls of estuarine mesocosms. Marine Ecology-Progress Series 155:1–15.

Cooke, G.D. 1967. The pattern of autotrophic succession in laboratory microcosms. BioScience 17:717–721.

Cooper, D.C. 1973. Enhancement of net primary productivity by herbivore grazing in aquatic laboratory microcosms. Limnology and Oceanography 18:31–37.

Copeland, B.J. 1965. Evidence for regulation of community metabolism in a marine ecosystem. Ecology 46:563–564.

Davis, W.J. 1993. Contamination of coastal versus open ocean surface waters: A brief meta-analysis. Marine Pollution Bulletin 26:128–134.

Dennison, W.C. 2007. Environmental problem solving in coastal ecosystems: A paradigm shift to sustainability. Estuarine Coastal and Shelf Science. 77(2):185–196.

Dewey, S.L. and F. deNoyelles. 1994. On the use of ecosystem stability measurements in ecological effects testing. Pages 605–625 in R.L. Graney, J.H. Kennedy and J.H. Rodgers, Jr. (eds.). Aquatic Mesocosm Studies in Ecological Risk Assessment. CRC Press, Boca Raton, FL.

Drake, J.A., G.R. Huxel and C.L. Hewitt. 1996. Microcosms as models for generating and testing community theory. Ecology 77:670–677.

Edmondson, W.T. and Y.H. Edmondson. 1947. Measurements of production in fertilized salt water. Journal of Marine Research 6:228–246.

Fee, E.J. and R.E. Hecky. 1992. Introduction to the northwest Ontario lake size series (NOLSS). Canadian Journal of Fisheries and Aquatic Sciences 49:2434–2444.

Frost, T.M., D.L. DeAngelis, S.M. Bartell, D.J. Hall and S.H. Hurlbert. 1988. Scale in the design and interpretation of aquatic community research. Pages 229–258 in S.R. Carpenter (ed.). Complex Interactions in Lake Communities. Springer-Verlag, New York.

Gamble, J.C. and J.M. Davies. 1982. Application of enclosures to the study of marine pelagic systems. Pages 25–48 in G.D. Grice and M.R. Reeve (eds.). Marine Mesocosms: Biological and Chemical Research in Experimental Ecosystems. Springer-Verlag, New York.

Gardner, R.H., W.M. Kemp, V.S. Kennedy and J.E. Petersen (eds.). 2001. Scaling Relations in Experimental Ecology. Columbia University Press, New York.

Giddings, J.M. and G.K. Eddlemon. 1977. The effects of microcosm size and substrate type on aquatic microcosm behavior and arsenic transport. Archives of Environmental Contamination and Toxicology 6:491–505.

Giesy, J.P., Jr. (ed.). 1980. Microcosms in Ecological Research. National Technical Information Service, Springfield, VA.

Goodyear, C.P., C.E. Boyd and R.J. Beyers. 1972. Relationships between primary productivity and mosquitofish (Gambusia affinis) production in large microcosms. Limnology and Oceanography 17:445–450.

Graney, R.L., J.H. Kennedy and J.H. Rodgers, Jr. (eds.). 1994. Aquatic Mesocosm Studies in Ecological Risk Assessment. CRC Press, Lewis Publishers, Boca Raton, FL.

Grice, G.D. 1984. Use of enclosures in studying stress on plankton communities. Pages 563–574 in H.H. White (ed.). Concepts in Marine Pollution Measurements. Maryland Sea Grant, College Park, MD.

Grice, G.D. and M.R. Reeve (eds.). 1982. Marine Mesocosms: Biological and Chemical Research in Experimental Ecosystems. Springer-Verlag, New York.

Grice, G.D., M.R. Reeve, P. Koeller and D.W. Menzel. 1977. The use of large volume, transparent, enclosed sea-surface water columns in the study of stress on plankton ecosystems. Helgoland Marine Research 30:118–133.

Harte, J., D. Levy, J. Rees and E. Saegebarth. 1980. Making microcosms an effective assessment tool. Pages 105–137 in J.P. Giesy, Jr. (ed.). Microcosms in Ecological Research. National Technical Information Service, Springfield, VA.

Have, A. 1990. Microslides as microcosms for the study of ciliate communities. Transactions of the American Microscopical Society 109:129–140.

Ives, A.R., J. Foufopoulos, E.D. Klopfer, J.L. Klug and T.M. Palmer. 1996. Bottle or big-scale studies: How do we do ecology? Ecology 77:681–685.

Kareiva, P. and M. Andersen. 1988. Spatial aspects of species interactions: The wedding of models and experiments. Pages 38–54 in A. Hastings (ed.). Community Ecology. Springer-Verlag, New York.

Kelly, J.R. and S.W. Nixon. 1984. Experimental studies of the effect of organic deposition on the metabolism of a coastal marine bottom community. Marine Ecology-Progress Series 17:157–169.

Kemmerer, A.J. 1968. A method to determine fertilization requirements of a small sport fishing lake. Transactions of the American Fisheries Society 97:425–428.

Kemp, W.M., M.R. Lewis, J.J. Cunningham, J.C. Stevenson and W.R. Boynton. 1980. Microcosms, macrophytes, and hierarchies: Environmental research in the Chesapeake Bay. Pages 911–936 in J.P. Giesy, Jr. (ed.). Microcosms in Ecological Research. National Technical Information Service, Springfield, VA.

Kemp, W.M., J.E. Petersen and R.H. Gardner. 2001. Scale-dependence and the problem of extrapolation: Implications for experimental and natural coastal ecosystems. Pages 3–57 in R.H. Gardner, W.M. Kemp, V.S. Kennedy and J.E. Petersen (eds.). Scaling Relations in Experimental Ecology. Columbia University Press, New York.

Kimball, K.D. and S.A. Levin. 1985. Limitations of laboratory bioassays: The need for ecosystem-level testing. BioScience 35:165–171.

Kitchens, W.M. 1979. Development of a salt marsh microecosystem. International Journal of Environmental Studies 13:109–118.

Krummel, J.R., R.H. Gardner, G. Sugihara, R.V. O'Neill and P.R. Coleman. 1987. Landscape patterns in a disturbed environment. Oikos 48:321–324.

Kuiper, J. 1982. Ecotoxicological experiments with marine plankton communities in plastic bags. Pages 181–193 in G.D. Grice and M.R. Reeve (eds.). Marine Mesocosms: Biological and Chemical Research in Experimental Ecosystems. Springer-Verlag, New York.

Lalli, C.M. (ed.). 1990. Enclosed Experimental Marine Ecosystems: A Review and Recommendations. Springer-Verlag, New York.

Lawton, J.H. 1998. Ecological experiments with model systems: The Ecotron facility in context. Pages 170–182 in W.J. Resetarits,Jr., and J. Bernardo (eds.). Experimental Ecology: Issues and Perspectives. Oxford University Press, New York.

Luckinbill, L.S. 1974. The effects of space and enrichment on a predator-prey system. Ecology 55:1142–1147.

Margalef, R. 1967. Laboratory analogues of estuarine plankton systems. Pages 515–521 in G. Lauff (ed.). Estuaries. American Association for the Advancement of Science, Washington, DC.

Maund, S.J., A. Peither, E.J. Taylor, I. Juttner, R. Beyerle-Pfnur, J.P. Lay and D. Pascoe. 1992. Toxicity of lindane to freshwater insect larvae in compartments of an experimental pond. Ecotoxicology and Environmental Safety 23:76–88.

McIntire, C.D. 1968. Structural characteristics of benthic algal communities in laboratory streams. Ecology 49:520–537.

Meentemeyer, V. and E.O. Box. 1987. Scale effects in landscape studies. Pages 15–34 in M.G. Turner (ed.). Landscape Heterogeneity and Disturbance. Springer-Verlag, New York.

Menzel, D.W. and J.H. Steele. 1978. The application of plastic enclosures to the study of pelagic marine biota. Rapports et Proces-Verbaux des Reunions – Conseil International pour L'Exploration de la Mer 173:7–12.

Myers, R.A. and B. Worm. 2003. Rapid world-wide depletion of predatory fish communities. Nature 432:280–283.

Naeem, S. and S.B. Li. 1998. Consumer species richness and autotrophic biomass. Ecology 79:2603–2615.

Nixon, S.W. 1969. A synthetic microcosm. Limnology and Oceanography 14:142–145.

Nixon, S.W., D. Alonso, M.E.Q. Pilson and B.A. Buckley. 1980. Turbulent mixing in aquatic mesocosms. Pages 818–849 in J.P. Giesy,Jr. (ed.). Microcosms in Ecological Research. National Technical Information Service, Springfield, VA.

Nixon, S.W., M.E.Q. Pilson, C.A. Oviatt, P. Donaghay, B. Sullivan, S. Seitzinger, D. Rudnick and J. Frithsen. 1984. Eutrophication of a coastal marine ecosystem – An experimental study using the MERL microcosms. Pages 105–135 in M.J.R. Fasham (ed.). Flows of Energy and Materials in Marine Ecosystems: Theory and Practice. Plenum Publishing, New York.

O'Neill, R.V. 1989. Perspectives in hierarchy and scale. Pages 140–156 in J. Roughgarden, R.M. May and S.A. Levin (eds.). Perspectives in Ecological Theory. Princeton University Press, Princeton, NJ.

O'Neill, R.V., D.L. DeAngelis, J.B. Waide and G.E. Allen. 1986. A Hierarchical Concept of Ecosystems. Princeton University Press, Princeton, NJ.

Odum, E.P. 1984. The mesocosm. BioScience 34:558–562.

Odum, H.T. 1951. The stability of the world strontium cycle. Science 114:407–411.

Odum, H.T. 1989. Experimental study of self-organization in estuarine ponds. Pages 291–340 in W.J. Mitsch and S.E. Jorgensen (eds.). Ecological Engineering: An Introduction to Ecotechnology. Wiley-Interscience, New York.

Odum, H.T. and C.M. Hoskin. 1957. Metabolism of a laboratory stream microcosm. Publications of the Institute of Marine Science, University of Texas 4:115–133.

Odum, H.T., W.L. Siler, R.J. Beyers and N. Armstrong. 1963. Experiments with engineering of marine ecosystems. Publications of the Institute of Marine Science, University of Texas 9:373–403.

Oviatt, C.A., S.W. Nixon, K.T. Perez and B. Buckley. 1979. On the season and nature of perturbations in microcosm experiments. Pages 143–164 in R.F. Dame (ed.). Marsh-Estuarine Systems Simulation. University of South Carolina Press, Columbia.

Petersen, J.E. 2001. Adding artificial feedback to a simple aquatic ecosystem: The cybernetic nature of ecosystems revisited. Oikos 94:533–547.

Petersen, J.E. and A. Hastings. 2001. Dimensional approaches to scaling experimental ecosystems: Designing mousetraps to catch elephants. American Naturalist 157:324–333.

Petersen, J.E., J.C. Cornwell and W.M. Kemp. 1999. Implicit scaling in the design of experimental aquatic ecosystems. Oikos 85:3–18.

Petersen, J.E., W.M. Kemp, R. Bartleson, W.R. Boynton, C.-C. Chen, J.C. Cornwell, R.H. Gardner, D.C.Hinkle, E.D. Houde, T.C. Malone,W.P. Mowitt,L. Murray, L.P. Sanford, J.C. Stevenson, K.L. Sundberg and S.E. Suttles. 2003. Multiscale experiments in coastal ecology: Improving realism and advancing theory. BioScience 53:1181–1197.

Peterson, D.L. and V.T. Parker (eds.). 1998. Ecological Scale: Theory and Applications. Columbia University Press, New York.

Pilson, M.E.Q. and S.W. Nixon. 1980. Marine microcosms in ecological research. Pages 724–741 in J.P. Giesy, Jr. (ed.). Microcosms in Ecological Research. National Technical Information Service, Springfield, VA.

Reed, C. 1978. Species diversity in aquatic microecosystems. Ecology 59:481–488.

Ringelberg, J. and K. Kersting. 1978. Properties of an aquatic microecosystem: I. General introduction to the prototypes. Archiv für Hydrobiologie 83:47–68.

Schindler, D.W. 1987. Detecting ecosystem responses to anthropogenic stress. Canadian Journal of Fisheries and Aquatic Sciences 44 (Supplement 1):6–25.

Schindler, D.W. 1998. Replication versus realism: The need for ecosystemscale experiments. Ecosystems 1:323–334.

Sheldon, R.W., A. Prakash and W.H. Sutcliffe, Jr. 1972. The size distribution of particles in the ocean. Limnology and Oceanography 17:327–340.

Smyly, W.J.P. 1976. Some effects of enclosure on the zooplankton in a small lake. Freshwater Biology 6:241–251.

Taub, F.B. 1969. A biological model of a freshwater community: A gnotobiotic ecosystem. Limnology and Oceanography 14:136–142.

Taub, F.B. 1997. Unique information contributed by multispecies systems: Examples from the standardized aquatic microcosm. Ecological Applications 7:1103–1110.

Taylor, D., S. Nixon, S. Granger and B. Buckley. 1995. Nutrient limitation and the eutrophication of coastal lagoons. Marine Ecology-Progress Series 127:235–244.

Taylor, E.J., S.J. Maund, D. Bennett and D. Pascoe. 1994. Effects of 3,4-dichloroaniline on the growth of two freshwater macroinvertebrates in a stream mesocosm. Ecotoxicology and Environmental Safety 29:80–85.

Threlkeld, S.T. 1994. Benthic-pelagic interactions in shallow water columns: An experimentalist's perspective. Hydrobiologia 275/276:293–300.

Tilman, D. 1989. Ecological experimentation: Strengths and conceptual problems. Pages 136–157 in G.E. Likens (ed.). Long-Term Studies in Ecology. Springer-Verlag, New York.

Turner, M.G. and R.H. Gardner. 1991. Quantitative methods in landscape ecology: An introduction. Pages 3–14 in M.G. Turner and R.H. Gardner (eds.). Quantitative Methods in Landscape Ecology. Springer-Verlag.

Twilley, R.R., W.M. Kemp, K.W. Staver, J.C. Stevenson and W.R. Boynton. 1985. Nutrient enrichment of estuarine submersed vascular plant communities. 1. Algal growth and effects on production of plants and associated communities. Marine Ecology-Progress Series 23:179–191.

Verdonschot, P.F.M. 1996. Oligochaetes and eutrophication; an experiment over four years in outdoor mesocosms. Hydrobiologia 334:169–183.

Verduin, J. 1969. Critique of research methods involving plastic bags in aquatic environments. Transactions of the American Fisheries Society 98:335–336.

Weber, P. 1994. Resistance to pesticides growing. Pages 92–93 in L.R. Brown, H. Kane and D.M. Roodman (eds.). Vital Signs 1994. W.W. Norton, New York.

Whittaker, R.H. 1961. Experiments with radiophosphorus tracer in aquarium microcosms. Ecological Monographs 31:157–188.

Wiens, J.A. 1989. Spatial scaling in ecology. Functional Ecology 3:385–397.

Williams, S.L. and W.H. Adey. 1983. *Thalassia testudinum* Banks ex Koenig seedling success in a coral reef microcosm. Aquatic Botany 16:181–188.

Designing Experimental Ecosystem Studies

Mesocosms are used for conducting controlled experiments at the level of whole ecosystems or ecological subsystems. Each mesocosm must be designed to answer the particular research question of interest.

There is no single best design that will suit all research objectives. However, experimental ecosystems should be constructed to include organisms and environmental conditions relevant for addressing specific research questions. Experience suggests that there are a host of issues, questions, and decisions that must be considered in the design of any enclosed experimental aquatic ecosystem. Our objective in this section is to provide a guide for those involved in designing and interpreting the results of mesocosm experiments.

Issues of scale (i.e. space, time, and ecological complexity) are a critical consideration in the design of mesocosm research, and thus scale provides a central theme for this book. The processes, organisms, and habitats under investigation determine the appropriate size, shape, and duration for the experimental ecosystem. Communities composed of small organisms can be studied in relatively small containers in relatively short-term experiments. Similarly, physiological processes, such as responses to light and nutrients, can be studied in short-duration experiments. Larger organisms and longer-term processes typically require longer-duration experiments.

General principles
J.E. Petersen and W.M. Kemp

The appropriate scale of research reflects choices between control, realism, and generality

Recently, aquatic ecologists have debated the appropriate scales for conducting ecosystem-level research. It is commonly recognized that the choices researchers make in any investigation reflect a balance (Fig. 1) between *control* (the ability to relate cause and effect, to replicate, and to repeat experiments), *realism* (the degree to which results accurately mimic nature), and *generality* (the breadth of different systems to which results are applicable). Smaller-scale systems are more generally controllable but less realistic, and larger systems are less controllable but may be more realistic (Kemp et al. 1980) Generality of results can be extended by conducting multiple experiments and using simulation models.

Ecosystem experiments can roughly be divided into two categories–*open field experiments* and *enclosure experiments*. Enclosure experiments (microcosms and mesocosms) have gained in popularity because they maximize opportunities for control. (Kemp et al. 1980, 2001)

In the last decade, a number of critics have suggested that the high degree of control afforded by mesocosm research comes at the expense of realism (Carpenter 1996; Schindler 1998; Roush 1995; Resetarits and Fauth 1998). Others have pointed out that issues of realism and other scaling problems are equally evident in whole ecosystem manipulations (Petersen et al. 2003). How, for example, does one extrapolate results from a small experimental lake to larger, deeper, more open, and more heterogeneous lake ecosystems that are the implicit focus of inference (Fee and Hecky 1992; Schindler and Scheuerell 2002)?

Researchers conduct open field experiments in Chesapeake Bay. Field research maximizes realism.

Realism   Generality

Control

Enclosure experiments enable a high degree of control, thereby allowing researchers to investigate causes and effects.

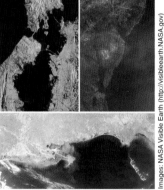

Generality refers to the range of systems to which experimental results can be applied. Top left: San Francisco Bay; Top right: Chesapeake Bay; Bottom: Gulf of Mexico.

Figure 1: To determine the appropriate scale of research, researchers must balance the competing goals of achieving control, realism, and generality.

Experimental ecosystems are simplifications and abstractions of nature

Consciously or not, researchers select a level of abstraction for their experimental system. The choices made have direct bearing on tradeoffs between control, realism, and generality. One can distinguish between *generic* and *ecosystem-specific* models, which represent the two extremes in this tradeoff (Fig. 2). Generic mesocosms are used to test broad theories that potentially apply to many different kinds of ecosystems. These systems tend to be small, highly artificial, have minimal physical and biological complexity, and are not designed to represent particular natural ecosystems. This is the ecological analog of using the nematode *Caenorhabditis elegans* as a model for studying development genetics in humans. Issues in ecology that have been successfully explored with generic mesocosms include succession development, predator-prey relations, stress, and stability versus diversity (Cooke 1967; Luckinbill 1973; Naeem and Li 1997). Because precise correspondence with particular ecosystems is not an objective, the researcher has considerable design flexibility in constructing generic models. The downside is that extrapolation from simplified, abstract systems to particular natural ecosystems is challenging (Øiestad 1990; Santschi et al. 1984).

Ecosystem-specific mesocosms are used to test hypotheses linked to particular types of ecosystems. This is the ecological analog of using chimps or human subjects to study human physiology and behavior. To achieve the higher degree of realism required, ecosystem-based mesocosms must incorporate the essential physical and biological features that control the dynamics in the systems that they represent. Various ecosystem-specific models have been constructed, ranging from coral reefs to rain forests (Luckett et al. 1996; Cohen and Tilman 1996). As the desired degree of specificity and desired level of realism increase, so does the complexity of engineering necessary to achieve realistic ecological conditions.

Obviously a continuum exists between generic and ecosystem-specific models, with tradeoffs analogous to those involved in modeling human physiology or genetics with use of nematodes, insects, rats, or chimps.

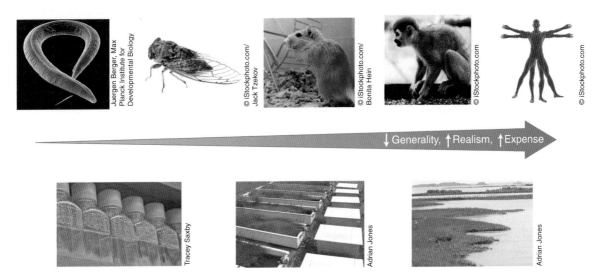

Figure 2: Tradeoffs between control, realism, and generality are evident in both organism and ecosystem level studies. For example, a nematode may serve as a general and inexpensive (though not very realistic) model of human physiology and behavior. Likewise, a flask of plankton can serve as a highly simplified model of an estuary. As the sophistication of the experimental system increases, generality tends to decrease while realism and expense tend to increase.

Scale decisions are crucial when designing experimental ecosystems

Empirical researchers can take two approaches to addressing issues of scale in their work. First, all researchers can strive to conduct *scale-sensitive experiments*. That is, scientists should use the best available theory and empirical work to explicitly consider and account for scale effects in both the design and interpretation of experiments. Second, ecologists who are interested in scale as a research question can engage in *multi-scale experiments* that manipulate attributes of time, space, or complexity in order to test specific hypotheses regarding the ecological impact of scale. Both scale-sensitive and multiscale experiments are prerequisites for improved mesocosm design and for the systematic extrapolation of information from experimental ecosystems to nature.

This book makes clear that scale is crucial to experimental design. The scale-related questions that follow and are linked in Table 1 should be considered by all researchers:

- How specific an ecological model is needed to answer a research question?
- What type of ecosystem will be simulated planktonic, pelagic-benthic, aquatic grass, or marsh?
- What resources are available in terms of funds, time, equipment, and support?
- What is the minimum system size, experimental duration, and ecological complexity necessary to answer thoroughly the motivating research question?
- How will the experimental system or systems address each of the following design issues: light source, mixing, temperature, water exchange, sediments, and source and introduction of organisms? (Fig. 3)

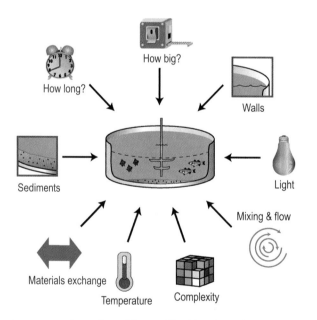

Figure 3: *The design choices that a researcher makes will inevitably affect the ecological dynamics that are observed. Scaling choices about system size, experiment duration, ecological complexity, and exchange rates are coupled with a variety of other design decisions including whether to clean walls, whether to use natural or artificial lighting, and how to mix the water column. It is important that the implications of these decisions be carefully considered.*

Design choices and implications

Table 1 *Researchers must consider various scale-related variables when designing experimental ecosystems.*

	Variable	Design decisions	Variables affected	Pages
	Size	Volume, depth, radius, surface area	Relative dominance of pelagic, benthic, and emergent producer communities, wall growth, temperature oscillations	50–58
	Time	Duration, timing of perturbation, sampling frequency	Ecological dynamics and life cycle of organism included in experiment, ability to detect seasonal and long term effects, influence of experimental artifacts	59–62
	Mixing	Vertical and horizontal mixing environment, mechanical mixing apparatus employed	Pelagic-benthic interactions, feeding rates and behaviour, access to nutrients, artifacts, and potential mortality associated with mechanical devices	63–74
	Materials exchange	Frequency, magnitude, chemical composition, biological composition (in both inflow and outflow)	Recolonization rates, flushing of planktonic organisms, selection for particular organisms and communities	75–81
	Light	Natural or artificial, light intensity, spectral properties	Primary productivity, producer community composition, water temperature	82–86
	Walls	Construction materials, cleaning frequency	Relative importance of periphyton community, light environment	87–90
	Temperature	Whether to control, how to control	Rate of biogeochemical activity, selection for particular organisms	91–92
	Sediments	Origin (intact core from nature or synthesized), particle size, organic matter content, organisms included	Pelagic-benthic interactions, seed banks, vascular plant growth, primary productivity	93–101
	Ecological complexity	Species and functional group diversity, number of habitats and biogeochemical environments included	Primary productivity, trophic dynamics, biogeochemical pathways, resistant and resilent stability	102–114

48 EXPERIMENTAL ECOSYSTEMS

Space and money constrain experimental ecosystem size, duration, and complexity

In scientific research, as in almost every other endeavor, resources are limited. Experimental ecosystem research is no exception, and the constraints of money, space, and time are always primary considerations. Smaller systems are cheap, take little space, and can be assembled and operated quickly. They offer opportunities for greater replication, more treatments, and more controllability, but sometimes at the expense of realism and complexity. Larger systems are expensive, take more space, and generally take longer to construct and operate. They are sometimes more realistic and can incorporate greater complexity, but achieving sufficient replication with multiple treatments can be a challenge. Therefore, a balance must be struck between realism, complexity, controllability, numbers of treatments, replication, and cost.

For example, a laboratory experiment with beakers may cost only a few thousand dollars to set up, operate for a short period of time, and analyze (Fig. 4). An experimental ecosystem facility on the scale of the MEERC Pelagic-Benthic facility may cost hundreds of thousands of dollars but will allow for experiments that may last for weeks at a time. A truly large-scale facility such as Biosphere 2 may be able to incorporate certain features of nature that would be impossible to incorporate in smaller systems, but costs millions of dollars to construct and to operate.

Bigger is not always better, however. The experience of very large systems such as Biosphere 2 demonstrates that increasing size, complexity, and expense does not inevitably lead to increasing realism. In some cases, depending largely on the questions asked, a microbial community in a flask may provide a more realistic model of nature than multiple biomes housed in an enormous closed structure.

Given a certain amount of space for experimental ecosystems, one can either achieve a high degree of replication with small systems or a low degree of replication with large systems. In MEERC's Pelagic-Benthic experiments, 12–15 intermediate-size (1 m^3) experimental ecosystems proved to be the most useful for questions related to the basic ecology of this type of ecosystem. They allowed multiple experiments or treatments, with multiple (usually three) replicates. Smaller systems were dominated by artifacts. Larger mesocosms were more realistic, especially for larger organisms, but were more difficult to operate. Finding an optimum size that best answers the research question while balancing the constraints of money, space, and time with the requirements of ecosystem complexity, realism, controllability, and repeatability, is an important task.

Thousands of dollars  Tens of thousands of dollars Millions of dollars

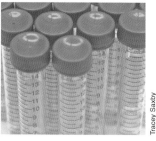  

Increasing time, space, complexity, and expense

Figure 4: Experiments may involve simple arrays of test-tubes (left), larger experimental ecosystems as in MEERC (middle), or expansive and complex facilities such as Biosphere 2 (right). Expense increases with all aspects of scale.

Spatial and temporal scaling
W.M. Kemp, J.E. Petersen, E.D. Houde, C.-C. Chen, J.C. Cornwell, and E.T. Porter

Spatial and temporal scales in mesocosms affect our ability to extrapolate from experiments to nature

The spatial and temporal scales of mesocosms and related experimental ecosystems regulate responses of system components to perturbations and other changes in external conditions. Therefore, a fundamental understanding of the scale-dependence of experimental ecosystem behavior is essential for quantitative extrapolation of mesocosm results to conditions in nature. The spatial and temporal scales of experimental coastal ecosystems studied in MEERC and other research are generally much smaller than the scales of the natural ecosystems to which one might wish to extrapolate research results.

Scaling theory predicts that certain patterns and processes only become evident as scale is increased beyond thresholds of extent (Wiens 2001). Furthermore, our own hypotheses and data indicate that scaling patterns tend to be non-linear (Kemp et al. 2001). So it is possible, for instance, that patterns that were determined

to be scale-dependent in MEERC mesocosm experiments become scale-independent at larger scales of natural systems (solid line in Fig. 5). Likewise, it is possible that relationships that one sees as scale-independent in mesocosms are functions of scale in larger natural ecosystems (dashed line in Fig. 5). Finally, it is possible that thresholds exist over which small changes in scale result in dramatic and possibly discontinuous changes in ecological dynamics. Given these possibilities, it is important that findings from multi-scale experiments be validated with data collected from a range of larger scale ecosystems in nature (Petersen et al. 2003).

In this section, results from MEERC and other research are used to develop scaling guidelines for designing experimental systems and conducting experimental research. This discussion considers spatial scales (size, shape) and temporal scales (duration, frequency) of experimental ecosystem studies.

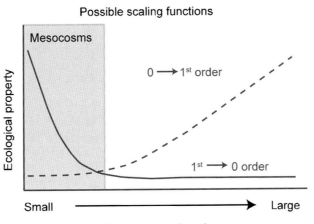

Figure 5: *Hypothetical responses of two distinct ecological properties to changes in the scales over which they are observed. Mesocosm scales (shaded region of graph) are inherently smaller than the scales of most natural systems. Trajectories shown indicate how different properties may be affected differently by changes in scale (Kemp et al. 2001; Petersen et al. 2003).*

Spatial scaling: Container depth, width, and shape constrain ecological properties of mesocosms

Aquatic mesocosms are characterized by clearly defined boundaries, surfaces, and dimensions (Fig. 6). The size and shape of cylindrical mesocosms can be described by two linear dimensions (depth, z, and width or radius, r) and two volumes (water and sediment [if present], where h is the height of sediments). In addition, there are three surface areas (air–water surface, sediment–water surface, mesocosm wall surface) that define the mesocosm's boundaries (Fig. 6). Light enters the mesocosms primarily at the air–water interface, which is also the site across which important gases such as oxygen, carbon dioxide and nitrogen are exchanged between experimental water and the overlying atmosphere. At the sediment–water interface, key solutes (e.g., oxygen, inorganic nutrients) move between water and sediment as a result of benthic biogeochemical processes. Also across this interface, particulate materials sink from the water column to the sediments and may be resuspended back into the water. The walls of aquatic mesocosms constitute a unique surface representative of hard substrates that typically occur in nature at much lower ratios of surface area to water volume. Walls constrain lateral exchanges of water and organisms between the container volume and surrounding region, and they provide a physical substrate for growth of attached organisms.

MEERC researchers examined 360 reports for data on mesocosm dimensions to explore patterns of shape, with an initial hypothesis that researchers may have implicitly tended toward use of experimental systems (Fig. 7) characterized either by constant depth ($z = C_1$), constant radius ($r = C_2$), or constant shape ($r/z = C_3$). Meta-analysis of reported dimensions of experimental systems revealed that published research results have been generated from mesocosms having remarkably similar shape (Petersen et al. 1999).

This shape-bias shared among researchers using enclosed experimental ecosystems is

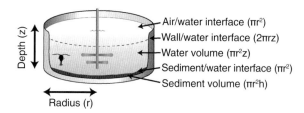

Figure 6: *Conceptual diagram showing key regions of biogeochemical activity associated with the spatial scales of a pelagic-benthic mesocosm. (h=sediment depth).*

somewhat disturbing. This is because the similarity in shape of aquatic mesocosms implies that there is no consistency in other important geometric properties of experimental systems. Specifically, for containers with constant shape, the relative importance (per water volume) of both wall area and horizontal surface area will tend to decrease with the size of the experimental system. Hence, the relative importance of wall artifacts such as periphyton growth and benthic processes such as nutrient regeneration will differ among experimental containers in proportion to their size (Chap. 1, Fig. 34).

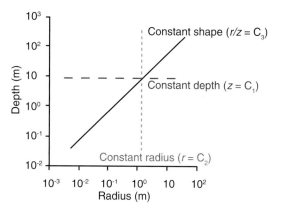

Figure 7: *Alternative simple relationships by which depth and radius of mesocosms might tend to be related. Constant radius: depth is varied but width (radius) is held constant. Constant depth: radius is varied but depth is held constant. Constant shape: depth and radius are varied in constant proportion to one another. Researchers tend to use containers with relatively constant shape (Petersen et al. 1999).*

Spatial scaling: Ecological effects of water column depth depend on light and nutrient conditions

In the MEERC pelagic-benthic experimental studies, fifteen cylindrical mesocosms were constructed with 5 distinct dimensions, 3 volumes, and 3 replicates per dimension (Fig. 8). These mesocosms were organized into three series: one with a constant depth (A, C and E tanks; depth = 1.0 m), one with constant shape (B, C, and D tanks; radius/depth = 0.56 m), and one with constant volume (E and D tanks = 10 m^3; A and B tanks = 0.1 m^3; Chap. 1, Fig. 41). These mesocosms with estuarine water and sediments were used to conduct a series of multiscale studies examining how container size and shape influence ecosystem behavior.

Scientists in MEERC developed a set of simple scaling hypotheses related to variations in depth of the upper mixed water column. Depth-scaling hypotheses started with the understanding that primary productivity in temperate coastal ecosystems often experiences a seasonal shift from light limitation in the spring to nutrient limitation in the summer. An important dimensional difference between these two limiting factors is that light energy is received on an areal basis (e.g., units of μmol photons m^{-2} s^{-1}) and is then absorbed as it travels down to deeper parts of the water column. In contrast, plankton experience nutrients on a volumetric basis (e.g., μmol m^{-3}), and concentration is relatively constant over depth in a well-mixed water column. These dimensional differences in nutrients and light suggest two simple depth-related scaling hypotheses that were tested in our mesocosm experiments.

Because aquatic ecosystems experience light on an areal basis, under purely light-limited conditions one might expect gross primary productivity (GPP) and related ecological variables to be constant among different depth systems when expressed on an areal basis: GPP_{Area} = C_1, (C_1 = a constant, units = g O_2 m^{-3} h^{-1}). In contrast, because biota contacts nutrients on a volumetric basis, under purely nutrient-limited conditions primary productivity should be constant when expressed per unit volume: GPP_{Vol} = C_2. In this nutrient-limited case, by definition productivity expressed per unit area should be directly proportional to depth: GPP_{Area} = $C_2 * z$.

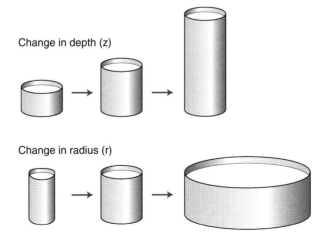

Change in depth (z)

Change in radius (r)

Figure 8*: Multi-scale studies in MEERC involved observing dynamics in a series of mesocosms that differed in container depth, radius, and shape.*

Spatial scaling: Water column depth regulates algal biomass and production

One set of MEERC multi-scale experiments illustrated how water column depth regulates gross primary production (GPP) of the integrated ecosystem by comparing the GPP (= daytime increases in O_2 minus nighttime decreases) for mesocosm systems of differing depth during spring (light-limited) and summer (nutrient-limited) conditions. Mean values for GPP normalized both per unit water volume and surface area generally followed the hypothesized relationships with spring GPP per area constant among systems but GPP per volume inversely related to depth (Fig. 9). For the nutrient-limited summer experiment, GPP per unit volume was relatively independent of water depth and, GPP per area increased with depth as predicted. These depth-scaling relationships derive from

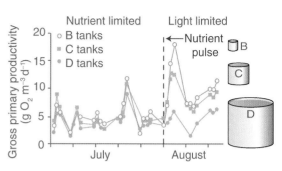

Figure 10: *Example time-series of GPP per water volume for three mesocosm types (B, C, D) of identical shape but different depth during nutrient limited summer before and after pulse addition of nutrients (Petersen et al. 2003).*

the fundamental differences in the way nutrient and light availability change with overall water column depth.

Further demonstration of the scaling relationship for nutrient-limited conditions is evident in temporal variations in GPP per unit volume for the three systems with the same shape but different depth (B, C, D) during summer (Fig. 10). Under these nutrient-limited conditions, GPP per unit volume was constant among the three mesocosm types despite differences in depth. Gross primary production expressed per unit volume exhibited the same temporal patterns of weekly variations in the different mesocosms until early August when a nutrient pulse was added to all systems. With this addition and the associated temporary transformation from nutrient-limited to light-limited conditions, GPP per volume increased inversely with water depth as predicted.

These experimental findings indicate the value of multi-scale experimental ecosystem studies in revealing fundamental scaling relationships. They reinforce the conclusion that the design and interpretation of mesocosm experiments must consider the water depth, nutrient conditions, and water clarity of the experimental system in relationship to these properties in the natural environment that the mesocosm is designed to represent.

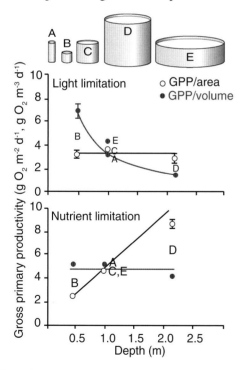

Figure 9: *Variations in gross primary productivity (GPP) in mesocosms of different depths. Mean GPP values (calculated both per water volume and per water surface area) and water column depth for (top panel) spring experiments under light-limited algal growth conditions and (bottom panel) summer experiments under nutrient-limited growth conditions.*

Spatial scaling: Enclosure depth affects primary productivity and zooplankton abundance

In situations where ecosystem production tends to be light-limited (e.g., temperate estuaries in spring), the abundance of zooplankton (the major plankton grazers in many natural and experimental coastal ecosystems) tends to vary with water depth in a relationship similar to that for the gross primary production (GPP). For the spring MEERC experiments, concentrations of dissolved inorganic nutrients (N, P, Si) were well above levels that limit primary production, such that availability of light was relatively more important as a control on algal growth. When mean values for both GPP and zooplankton biomass (both expressed per m^3 water volume) in the mesocosms were combined with similar spring data for stations in the mesohaline regions of Chesapeake Bay and one of its major tributary systems (Patuxent River estuary), there were parallel trends with increasing depth (Fig. 11).

Initially, a statistically significant negative exponential equation was developed using only the mesocosm data (within the shaded area to the left). However, the regressions were unchanged when Bay data were included, suggesting that mesocosm scaling relationships could be extrapolated directly to conditions in nature. In this case, however, all of the significant changes in GPP and zooplankton occur within the mesocosm range of depths, such that changes in these variables become essentially independent of depth when depth exceeds 2 or 3 m. This threshold point for depth-dependence likely varies with water clarity, becoming deeper in clearer water columns.

These results also imply that, under resource-limited growth conditions, scaling relationships may propagate from lower to higher trophic levels. This is essentially a situation of *bottom-up control* (i.e., resource limitation) on organism

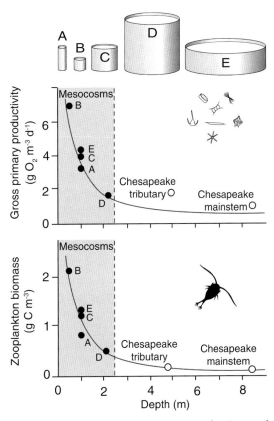

Figure 11: Variations in primary production and zooplankton biomass with changes in water column depth for five experimental and two natural estuarine ecosystems with similar salinity. Experimental ecosystems have five different sizes or shapes and the estuarine sites are in the mainstem and a tributary of Chesapeake Bay. Data are mean values for gross primary productivity (GPP per unit water volume) measured from changes in dissolved oxygen concentration (Kemp and Boynton 1984; Kemp et al. 2001; Roman unpublished).

growth and production. These depth-scaling relationships might be less evident under conditions of *top-down control* (i.e., limitation via consumption) on production resulting from externally induced changes in consumer abundance.

Spatial scaling: Enclosure depth affects nutrient recycling but does not simulate natural ecosystems well

For both gross primary productivity and zooplankton biomass, scaling coefficients derived from mesocosm data alone were almost identical to those derived by combining data from mesocosms and natural ecosystems (previous page). It is worth noting, however, that GPP data for the Patuxent estuary fall some distance from the fitted equation. Had data for other natural ecosystems been included, there would likely be considerable scatter around the regression because many factors other than depth vary among estuaries. Indeed, in many ways this is the point; a great strength of multi-scale mesocosm experiments is that these other factors, such as nutrient loading, can be held constant so that scaling effects can be isolated successfully. This comparison suggests that, at least for some variables and under some circumstances, scaling relationships revealed in mesocosms are robust and can be directly extrapolated to nature.

For other variables, however, the potential value of mesocosms for revealing fundamental scaling relationships and the potential for direct extrapolation may be more complicated. For instance, benthic recycling of dissolved inorganic nitrogen was expected to conform to the depth-scale hypothesis. In this case, although an inverse equation [$Y = N_1 (z)^{-1}$, where Y is the ratio of N recycling to N input and N_1 is a coefficient] neatly fitted data from the mesocosms (Fig. 12), the extrapolation of this equation (dashed line beyond mesocosm scales) did a poor job of predicting data gathered for larger natural ecosystems. The same inverse equation with a different coefficient did, however, fit the data from five field sites (dashed line).

There are a number of explanations for why scaling relationships for relative regeneration of dissolved inorganic nitrogen might differ between mesocosms and nature. For example, although the horizontally rotated impellers in the experimental systems produced realistic

mixing in the water column, the bottom shear velocity at the sediment-water interface was unrealistically low (≈ 0.1 cm s^{-1}). This enhances key benthic processes such as microphytobenthic production and nutrient uptake (Porter et al. 2004b); unrealistically low shear at the bottom interface would thus tend to inhibit nutrient regeneration rates. Moreover, in benthic-pelagic coupling experiments in MEERC a moderate bottom shear velocity of 0.6 cm s^{-1} resulted in microphytobenthos erosion and affected subsequent nutrient regeneration compared to experiments in tanks with the same water column mixing but low bottom shear (Porter et al. 2004b).

The important lesson is that although multi-scale experiments can be used to identify valid scaling relationships in some cases, unrealistic biological and physical conditions within mesocosm studies (i.e., an inadequate representation of physical complexity) can sometimes distort ecological dynamics and result in erroneous conclusions when extrapolating results from mesocosms to nature.

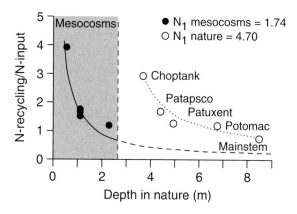

Figure 12*: Relationships between relative rates of benthic nitrogen recycling (as a fraction of N input) and depth in mesocosms and in large natural estuarine tributaries and the mainstem of Chesapeake Bay (Mesocosm data from Cornwell unpublished; field data from Boynton et al. 1995; Kemp et al. 2001; Petersen et al. 2003).*

Spatial scaling: Water depth and light availability control benthic microalgal production

Benthic primary production by microalgae tends to vary with water column depth simply because of depth-dependent variations in light levels reaching the sediment surface. Figure 13 demonstrates how depth regulates benthic primary production (BPP) by comparing estimated BPP for three experimental mesocosms having different water column depths (B, C, and D, with depths of 0.5, 1.0, 2.2 m) with BPP rate measurements made in the mesohaline region of Chesapeake Bay.

When rates of BPP are plotted versus water column depth for the mesocosms and the field sites, there are two parallel but different patterns of exponentially decreasing rates with depth (Fig. 13, top panel). It might be anticipated that this difference is simply related to differences in light regimes in the experimental and natural

environments. Although the shapes of the respective relationships between BPP and light at the sediment surface are similar for observations in the two environments, mesocosm rates at a given light level were consistently lower (Fig. 13, bottom panel). Presumably, differences between relationships of BPP versus light between mesocosms and the Bay are attributable to differences of nutrient availability or other factors between the experimental and natural habitats.

The presence of benthic microalgal communities can also affect nutrient recycling processes in experimental ecosystems because algal uptake of nutrients is proportional to light reaching the sediment, which is in turn related to water column depth. Recycling effluxes of dissolved inorganic nitrogen from sediments of shallow mesocosms with low bottom shear velocity decline significantly in the light because of assimilation by benthic microalgae (Fig. 14).

Thus, researchers must consider differences in habitat conditions when attempting to extend results from enclosed experiments to conditions in nature. On the other hand, experimental planktologists must realize that benthic communities will tend to develop at the bottom surface of all containers, and that these chamber bottoms may be sites of active organic production and consumption as well as nutrient cycling processes.

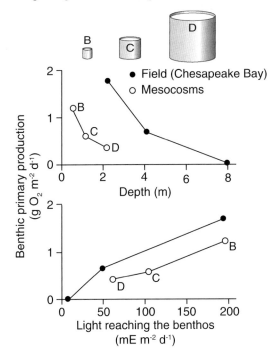

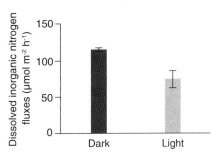

Figure 13: Benthic primary production versus water depth in mesocosms and Chesapeake Bay (top panel). Benthic primary production versus light intensity at the sediment surface in mesocosms and Chesapeake Bay (bottom panel) Computed from unpublished data of C.-C. Chen and Kemp 2004 for mesocosms and W. M. Kemp et al. 2005 for Chesapeake Bay.

Figure 14: Fluxes of dissolved inorganic nitrogen (nitrate + nitrite + ammonium) from sediments to overlying water under dark and light conditions in shallow experimental ecosystems with healthy benthic microalgal communities (Cornwell unpublished).

Spatial scaling: Wall artifacts vary with width of experimental ecosystems

Whereas differences in water column depth may lead to fundamental differences in ecological characteristics in experimental and natural systems, differences in mesocosm width may induce different levels of experimental artifacts. One potentially important artifact is growth of periphyton communities (including bacteria, algae, protozoa, and small metazoans) on mesocosm walls. Because walls are not normal habitats in most natural planktonic or benthic subsystems, growth of periphyton communities on container surfaces is usually considered a non-representative artifact that complicates extrapolation of results to conditions in nature. Experiments were conducted in MEERC pelagic-benthic mesocosms with tanks of different radius to quantify the artifacts of scale associated with this wall growth. If the relative contribution of wall productivity to total system productivity was proportional to the ratio of wall area to water volume ($A_w{:}V$), then it would also be inversely proportional to radius for a cylindrical container ($A_w{:}V = 2\pi r \times z/\pi r^2 \times z = 2/r$). Thus, the relative contribution of walls to total gross primary productivity among systems of different radius would be $\mathrm{GPP}_{wall}/\mathrm{GPP}_{total} = C_3/r$, where C_3 is a constant and r is the container radius.

In experiments without routine removal of material from container walls, periphyton biomass (normalized per water volume) was inversely proportional to mesocosm radius (Fig. 15). Similar relationships were seen for periphyton GPP and periphyton uptake of dissolved inorganic nitrogen (Chen et al. 1997, 2000). Hence, the relative contribution of periphyton communities to total biomass and ecological processes in mesocosms tended to decrease with container radius. These observations led to the hypothesis that the ratio of $\mathrm{GPP}_{wall}/\mathrm{GPP}_{total}$ is related to the inverse of container radius (r^{-1}). Although this ratio did decline with container radius, the exact relationship did not conform perfectly with the hypothesized model (Fig. 16). When wall productivity was expressed per unit wall surface area, periphyton were more robust in wider systems. This trend may be related to self-limitation for light or nutrients or both, with denser periphyton growth or more intense grazing pressure in narrower tanks (Chen et al. 1997).

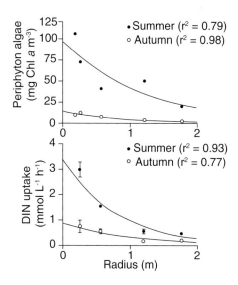

Figure 15: Variations in periphyton community properties with increasing container radius, including algal biomass (top panel) and uptake of dissolved inorganic nitrogen (bottom panel) both expressed per unit water column volume (Chen et al. 1997, 2000).

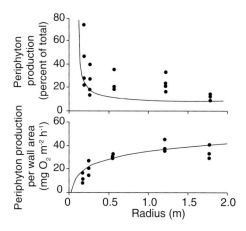

Figure 16: Changes in relative contribution of wall periphyton to total gross primary productivity (top panel) and in productivity of wall periphyton per unit of wall area (bottom panel) with mesocosm radius. The curved lines through the data are a least-squares fit using the hypothesized inverse relationship between relative productivity and tank radius $\mathrm{GPP}_{wall}/\mathrm{GPP}_{total} = C_3/r$ (top panel), and a best-fit line (bottom panel); (Chen et al. 1997; Petersen et al. 2003).

Spatial scaling: Mesocosm size constrains the behavior and physiology of mobile animals

The width of an experimental ecosystem also affects the behavior and physiology of fish and other swimming animals contained in that ecosystem. The frequency of encounters that a randomly swimming fish might have with container walls would be proportional to the ratio of wall area (A_w) to water volume (V), which is in turn proportional to container radius ($r = 2V/A_w$). More encounters with walls probably leads to more uncharacteristic animal behavior and associated physiological stress (Heath and Houde 2001).

A series of experiments was conducted in pelagic-benthic mesocosms (Mowitt et al. 2006). using the zooplantivorous bay anchovy (*Anchoa mitchilli*). In 43-day experiments, there was a consistently increasing rate of anchovy growth as the width of experimental containers increased (Fig. 17; Mowitt et al. 2006). At widths exceeding 1 m diameter, growth rates began to approach those observed in the natural estuarine environment (Chesapeake Bay).

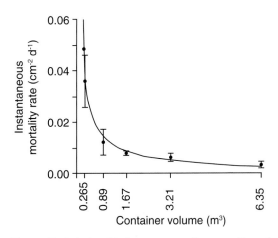

Figure 18: *Relationship between mortality rate of larval fish (capelin,* Mallotus villosus) *subjected to predation by gelatinous zooplankton* (Aurelia aurita) *at standardized size and density and volume of experimental container. Mean and standard deviation are presented for each enclosure volume (de Lafontaine and Leggett 1987).*

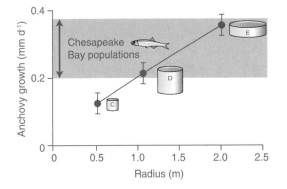

Figure 17: *Variations in growth rate (mean ± SE) of bay anchovy* (Anchoa mitchilli) *with increasing radius of experimental containers. Growth rates were measured as an increase in standard length of fish during period of captivity in container (Mowitt et al. 2006).*

Although precise mechanisms underlying this pattern are yet unknown, they presumably arise from altered swimming and feeding behavior of the bay anchovy, which may increase respiration and decrease prey capture.

Similar studies of predator-prey interactions have revealed aberrant behavior of the enclosed organisms in experiments conducted in small containers. For gelatinous zooplankton feeding on fish larvae, a steep inverse relationship between predation (prey mortality) rates and container volume was observed for mesocosms less than about 2 m³ volume (Fig. 18). Conversely, other studies have reported that predatory pelagic animals appear to have threshold minimum container sizes beyond which growth rates are unaffected by container walls. This critical container volume tends to vary directly with both size and density of the consumer animals (de Lafontaine and Leggett 1987).

Spatial scaling: Relative variance among and within replicate experimental ecosystems is affected by container size

In addition to the average characteristics of MEERC's experimental systems varying with size, the relative variability (variance/mean) of those properties also changes with container size. The question of variance scaling can be considered in two contexts: (1) variance among replicate experimental systems and (2) variance within a given experimental system.

In the first case, researchers observed that the relative variability in phytoplankton characteristics among replicate mesocosms appears to decline with size. In fact, relative variance decreased exponentially with mesocosm length scale (Fig. 19, top panel). Variance among replicate systems was hypothesized to be controlled largely by processes occurring at container surfaces: periphyton on walls, benthic communities at the sediment surface, and neuston (minute organisms at the air–water interface). Therefore, a length-of-edge index, the square root of the sum of these three surface areas, was used for analysis (Fig. 19). Although there is surprisingly little other published information to confirm the generality of this pattern, it appears that decisions by experimentalists on the number of replicates needed for studies in mesocosms of different size are consistent with this finding. Indeed, a review of the mesocosm literature revealed an inverse relationship between size of experimental container and number of replicates used by researchers. This may reflect an understanding that larger systems are more stable and more predictable, leading to lower variability among replicates. However, this trend may be driven more by logistical considerations, including the expense of building, housing, and maintaining many replicates for larger experimental systems.

The second kind of variance-scaling involves how within-the-system variability of ecological properties changes with the size of experimental containers. In this case, MEERC researchers found that relative variability in copepod abundance tended to increase with container size (size is measured again with the length-of-edge index). In general, within-system variance tended to increase with container size (Fig. 19, bottom panel). It is likely that this relationship emerges because larger systems have more space (particularly, near edges) within which patchiness can develop. It also appears that this trend becomes more pronounced with time during the experiment, such that this scaling effect is exacerbated by experimental duration.

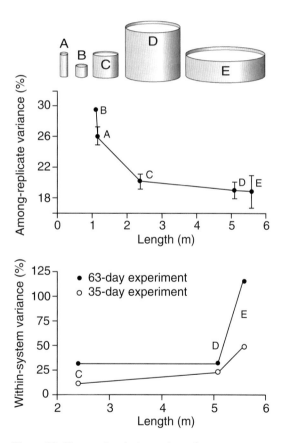

Figure 19*: Changes in relative variance for measurements of phytoplankton abundance among replicate experimental systems with changes in container size (top panel; Chen et al. 2000) and changes in relative variance for zooplankton abundance within experimental systems with changes in container size after 35 and 63 days of experiments (bottom panel; Kemp et al. 2001; Petersen et al. 2003).*

Temporal scaling: Primary producer abundance in experimental ecosystems characteristically increases, then declines

Previous investigators (Dudzik et al. 1979) have reported a tendency of experimental planktonic communities to exhibit a characteristic temporal pattern with an initial pulse (bloom) of phytoplankton growth followed by a decline and extended period of relatively low algal biomass (bust). The duration of the bloom period tends to be 1–2 weeks whereas the low biomass period is protracted. This pattern presents a potential problem for conducting investigations with plankton communities because the experimental duration needs to be limited in time so as to avoid extension across the two sequential regimes (bloom and bust). If an experimental manipulation extends across the two regimes, it will be difficult to separate treatment effects statistically from these background changes in control and treated systems.

Similar temporal patterns occurred in marsh mesocosms containing emergent vascular plants, but at different scales. These mesocosms, maintained continuously for several years, exhibited exceptional growth of marsh plants in the first year, followed by a gradual decline over the subsequent 4 years.

When the phytoplankton and marsh plant biomass data are presented versus a dimensionless time-scale based on the biomass turnover-time of the plants, the patterns are virtually identical (Fig. 20, top and middle panels). Here, turnover time is the time required for plant growth to replace existing biomass, and the dimensionless "relative time" is defined as (clock time)/(turnover time), where turnover times were taken as 1.4 days for algae and 30 days for marsh plants. For both primary producers, at high and low diversity, and in experimental containers of all sizes and shapes, peak biomass occurred after 2–4 relative turnovers. This sequence was hypothesized to be largely attributable to depletion of the initial nutrient supply. This idea was tested by comparing time-series of phytoplankton biomass in control containers with those receiving daily high and

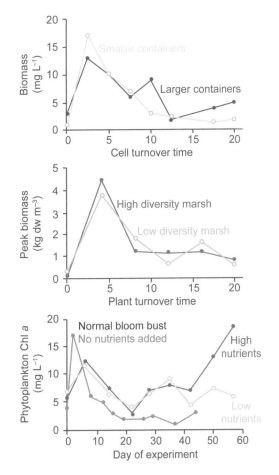

Figure 20*: Typical temporal patterns of variations in phytoplankton (top panel) and marsh plant biomass (middle panel) in mesocosms over relative time (clock time/ plant turnover time) for phytoplankton in smaller (0.1 m^3) or larger (10 m^3) containers (top panel) and for marsh plants in containers with one dominant plant species or five co-dominants (middle panel). The daily addition of nutrients (N + P + Si) at low or high rates to phytoplankton systems eliminated this bloom-bust pattern (bottom panel).*

low rate additions of nutrients (N, Si, P at 16:16:1 proportions). In fact, added nutrients resulted in the removal of the bloom-bust cycle, with algal biomass maintained throughout the 8-week experimental duration (Fig. 20, bottom panel). Thus, it appears that these bloom-bust cycles can be eliminated with continual nutrient addition or, perhaps, by including an adequate nutrient repository (e.g., sediments) in a mesocosm.

Temporal scaling: Zooplankton abundance in experimental ecosystems characteristically lags phytoplankton abundance

As with phytoplankton growing in experimental ecosystems, temporal trends in zooplankton biomass in mesocosms tend to follow characteristic cycles of growth and decline. Typical examples of these cycles are depicted for mesocosms of five different size and shapes during spring and for comparable regions of Chesapeake Bay (Fig. 21). Zooplankton assemblages, which are dominated by the copepod *Acartia tonsa*, exhibit peak biomass values that lag behind phytoplankton blooms by about 2 weeks, with zooplankton values being generally higher in the smaller mesocosms (A, B, and C in Fig. 21), possibly revealing a scaling effect. Whereas zooplankton and phytoplankton biomass levels were similar to each other in mesocosms, in the natural estuary zooplankton biomass is only 10–20% of that of their phytoplankton prey. In estuaries such as Chesapeake Bay, zooplankton biomass tends to be controlled more by predators than by food supply (Roman et al. 2005).

In typical experimental ecosystems, however, most planktivorous predators are excluded, thereby allowing initial explosive zooplankton growth followed by declining growth with diminishing food supplies.

The coincidence of declining phytoplankton and increasing zooplankton in mesocosms suggests that zooplankton grazing contributes to algal losses. In fact, nutrient enrichment experiments resulted in significant increases in zooplankton biomass with little change in phytoplankton abundance, indicating food-limited conditions for zooplankton (Roman unpublished). Under the extreme food limitation that typifies conditions in many experimental mesocosms, adult copepods will also prey heavily on their own juveniles. This kind of cannibalistic behavior appears to be much less common in natural estuaries where adult copepods are controlled by their predators (Roman unpublished).

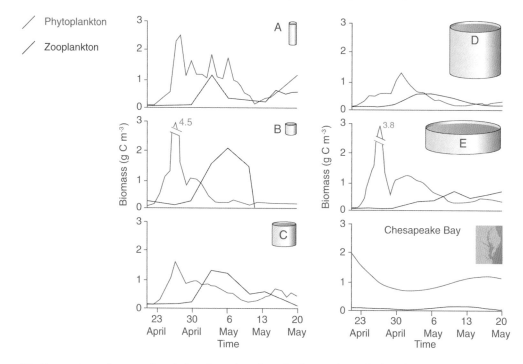

Figure 21*: Comparison of temporal variations in spring phytoplankton and zooplankton biomass levels in experimental ecosystems of different size (ranging from 0.1 m³ to 10 m³) and in the mesohaline region of Chesapeake Bay. (Mesocosm data from Roman unpublished; Field data are adapted from Brownlee and Jacobs 1987 and Harding et al. 2002).*

Temporal scaling: Extended experiments lead to depleted nutrients and food, and subsequent reduced fish growth

Earlier (p. 57), an example from experiments with planktivorous (zooplankton-eating) bay anchovy, *Anchoa mitchilli*, were used to illustrate how the size of experimental containers can affect the behavior and growth of captive fish. Although these results were from a relatively short duration (43-day) experiment, a longer (90-day) experiment produced a similar pattern of increasing fish growth with increasing container radius (Fig. 22). In this longer study, however, anchovy growth rates were significantly lower than those in the shorter experiment, and all rates from the longer study were well below the range of values measured in the estuary. Given the tendency for zooplankton biomass in experimental ecosystems to decline after several weeks without supplemental nutrient additions, (p. 59), these results are not surprising. Based on observed zooplankton abundances,

calculations of anchovy bioenergetic balance (i.e., consumption minus excretion minus respiration) for these two experiments revealed that growth of fish in the longer duration study was, in fact, acutely limited by food availability. To minimize stress on experimental fish, growth rates had been calculated for individual fish only at the end of these experiments by comparing initial size (and inferred weight) with measured size (and weight) at the end of the experiment. It was assumed that during their first 43 days in captivity, the fish in the 90-day experiment were probably growing at rates similar to those measured in the shorter study; however, after 90 days with limited food supplies, some individuals actually shrank (Mowitt 1999). Thus, duration of study in experimental estuarine ecosystems can substantially affect the observed productivity, standing stocks, feeding rates, and growth of contained organisms.

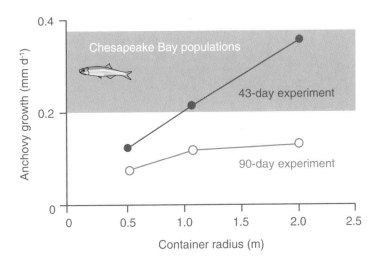

Figure 22: Mean (± SE) growth of bay anchovy, Anchoa mitchilli, *versus diameter of experimental ecosystems for similar studies of 43- and 90-day duration. Growth rates are calculated comparing length of fish at the beginning and end of the experiments (Mowitt 1999).*

Temporal (and spatial) scaling: Extended experiments lead to depleted nutrients, food, and subsequent reduced oyster growth

The previous example illustrated how the size and shape of experimental containers and the duration of an experiment can substantially alter the growth of planktivorous fish. This example demonstrates a similar pattern for benthic filter-feeding oysters (*Crassostrea virginica*) when experiments are conducted without continuous nutrient additions and without resuspension of bottom sediments. These kinds of experiments are commonly used to study benthic filtration effects on plankton in aquatic ecosystems.

In these experiments, the larger mesocosms (1.0 m³, C tanks) consistently had more phytoplankton biomass regardless of whether or not oysters were present. The smaller (0.1 m³, A tanks) narrower containers (Fig. 23; note differences in *Y*-axis scale between panels) had lower light penetration through the water column, thereby supporting less algal production. As previously discussed, phytoplankton biomass exhibited an initial bloom during the first week, followed by a gradual decline in abundance over the subsequent 4 weeks (Porter et al. 2004a; Porter 1999).

In this experiment, the filter-feeding rates were scaled to container volume in the two mesocosm types by adding 120 juvenile oysters to the 1 m³ mesocosms and 12 to the 0.1 m³ systems. Within a week of the initiation of the experiment, oyster filtration had essentially depleted phytoplankton standing stocks in all experimental ecosystems, with chlorophyll-*a* values below 1 µg L⁻¹ in the smaller systems and generally less than 5 µg L⁻¹ in the larger containers. Although chlorophyll *a* levels declined in the mesocosms without oysters as well, they remained significantly higher (2-6 µg L⁻¹ in the smaller systems and 10-40 µg L⁻¹ in the larger containers).

At these experimental animal densities and container volumes, oysters were probably food-limited for much of the experiment, especially in the small mesocosms. Consequently, over the 4-week-long experiments, oysters grew significantly better in the larger mesocosms with higher food abundance than in the smaller systems (Fig. 24; Porter et al. 2004a; Porter 1999). For a shorter experimental duration, the difference might not have been so large; however, light limitation in the smaller, narrow tanks may have created an inherently lower phytoplankton system.

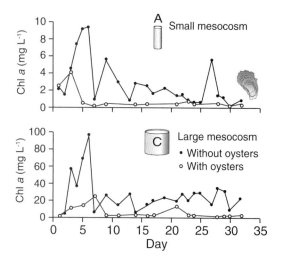

Figure 23: *Time-course changes in phytoplankton biomass (chl* a*) in smaller (0.1 m³, upper panel) and larger (1.0 m³, lower panel) mesocosms with and without benthic filter-feeding oyster populations. Notice the difference in Y-axis scale for the two experimental systems.*

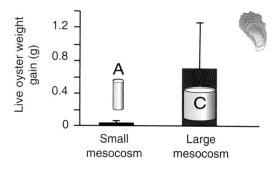

Figure 24: *Difference in weight gain (mean ± standard deviation) by oysters grown in smaller (0.1 m³) and in larger (1.0 m³) mesocosms. n=10 in the larger mesocosms, n=7 in the smaller mesocosms.*

Physical factors: Mixing and flow
L.P. Sanford, S.E. Suttles, and E.T. Porter

Mixing and flow influence all scales of natural coastal ecosystem processes

Flow may be defined as the net movement of a fluid (or a dissolved or suspended constituent in the fluid) through a cross-section, whereas *mixing* may be thought of as exchange of a fluid or its constituents between adjacent volumes with no net directional movement. Mixing and flow are important aspects of aquatic ecosystems from the largest to the smallest of scales.

At global scales, the flows of the atmosphere and the ocean are the major engines that drive heat and nutrient fluxes through the earth's ecosystem. In the North Atlantic, the Gulf Stream carries tropical heat northward and gives Europe its temperate climate. Also, mixing of water masses across the Gulf Stream can have significant effects on the oceanic ecosystems that it separates. Movement of the warm core rings of water from the ocean side of the Gulf Stream to the continental shelf side introduces significant amounts of water low in nutrients and algae into shelf environments (Fig. 25, upper left).

At intermediate (meso) scales, mixing and flow are crucial in estuaries and coastal waters where fresh and saltwater interact. Flow and mixing of water in an estuary are intricately linked processes that determine the estuary's water exchange rate, and in so doing these processes play a major role in estuarine productivity.

At very small scales, microscopic organisms are influenced by relative motion of the fluid (shear) that is directly related to mixing intensity (Fig. 25, upper right). This shear renews nutrient and food supplies, affects contact between predators and prey, and may be a source of physical stress at high levels.

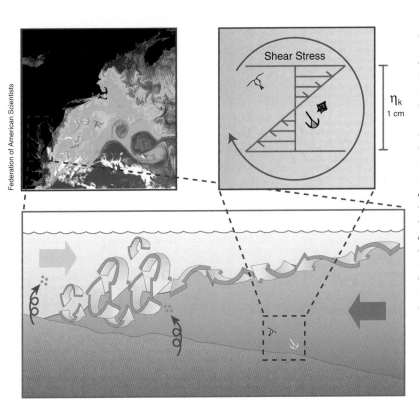

Figure 25: Illustration of the importance of mixing and flow across all scales in coastal ecosystems (see text). The upper left panel is an enhanced satellite image of ocean color in the North Atlantic Ocean off the U.S. east coast showing low chlorophyll, warm core rings from the Gulf Stream at large scales (km); the lower panel is an idealized cross-section of a partially mixed estuary illustrating mixing processes at intermediate scales (m); and the upper right panel is a schematic diagram of the smallest turbulent eddy illustrating the mixing and flow environment at very small scales (cm or less). (η_k = Kolmogorov microscale; see next page)

Mixing is accomplished through turbulent eddies

For very slow flows at very small scales, mixing is accomplished by molecular processes. Almost all natural flows are turbulent, however, such that most mixing in nature is through the exchange of larger blobs of fluid carried by turbulent eddies. The turbulent energy spectrum (Fig. 26) helps to illustrate how turbulent energy is transferred from the large scales at which it is generated towards ever smaller scales. In doing so, this turbulent energy "cascade" mixes adjacent fluids (and their constituents) down to scales at which molecular processes can finish smoothing out gradients. The spectrum shows the distribution of velocity variance across wave number (inverse eddy size). Low wave numbers represent low frequency, large-scale processes and higher wave numbers correspond to shorter time and space scales.

Energy is highest at the low wave numbers that correspond to the largest eddies. These large eddies break down into smaller and smaller eddies, "cascading" their energy towards higher wave numbers. The range of intermediate wave numbers is called the inertial-subrange, and is characterized by a $-5/3$ slope on a log-log plot.

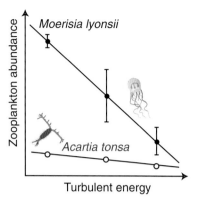

Figure 27: *Relationships between the abundance of* Moerisia lyonsii *and* Acartia tonsa, *and the turbulent energy dissipation rate (ε) in the $^{-5}/_3$ mixing treatments (Petersen et al. 1998).*

The velocity spectrum starts to decrease faster than the $^{-5}/_3$ slope at a wave number slightly below that of the smallest turbulent eddy (Kolmogorov mircoscale, η_k). This represents a transition to a range where the energy is dissipated by molecular viscosity. Shear still exists at scales smaller than the Kolmogorov microscale, but the turbulent spatial structure disappears and gradients are uniform. The same is true for microscale distributions of nutrients, particles, salt, and temperature, except that their smallest scales (the Batchelor scales) are much smaller because their diffusivities are much smaller than the viscosity of water.

Many past mesocosm experiments suffered from a lack of mixing, but too much mixing is undesirable as well. For example, too much mixing can affect the structure of a pelagic ecosystem. As mixing energy is increased above the level to which organisms are acclimated in the environment, many organisms are damaged. In a MEERC pelagic-benthic experiment where mixing energy, as measured by the turbulent energy dissipation rate (ε) was a treatment (Petersen et al. 1998), systems that were mixed at relatively high energies had their copepod populations decline slightly and gelatinous zooplankton populations decline precipitously (Fig. 27).

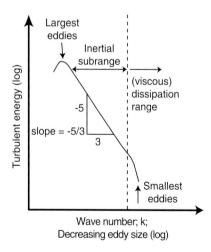

Figure 26: *Idealized velocity spectrum plot showing the cascade of turbulent kinetic energy, characterized by the −5/3 slope part of the curve on the log-log plot, from large eddies to the smallest turbulent eddies (Kolmogorov mircoscale, η_k). Wave number, k, corresponds to the inverse of eddy size.*

The relative importance of mixing and flow changes in pelagic and benthic environments

Mixing is biologically more important than flow in a pelagic environment and flow is more important than mixing in a benthic environment, although both must be considered in each environment (Fig. 28). Plankton in a pelagic ecosystem move with the flow of the water, so the speed of the flow itself is relatively unimportant. What matters to planktonic organisms is access to dissolved nutrients and oxygen, their encounter rates with prey or predators, and physical disturbance by turbulent mixing energy. In other words, mixing is the dominant physical influence on any given volume of the pelagic ecosystem. By contrast, many benthic organisms are attached to the bottom and are much more affected by flow. Water flow at moderate rates transports food, nutrients, and oxygen to bottom animals and plants, whereas high flow tends to exert direct physical stress on exposed organisms. Mixing is also important to the benthos because it determines the rate of exchange between the

very near bottom environment and the interior of the water column. However, near-bottom mixing is directly proportional to near-bottom flow because turbulent mixing is generated by flow over the bottom.

Turbulent mixing generated by flow over the bottom is affected by the roughness of the bottom, which may be due to sediment bedforms or benthic flora or fauna that protrude into the flow. When the bottom is hydraulically smooth (in the absence of significant bed roughness), there is a diffusive boundary layer just above the sediment-water interface where molecular diffusion dominates. The thickness of this layer is an important control on fluxes of dissolved substances between sediments and the overlying water column. The thickness of the diffusive boundary layer is inversely proportional to the strength of the flow and the turbulence it generates, so higher flows result in thinner diffusive boundary layers and less diffusive resistance.

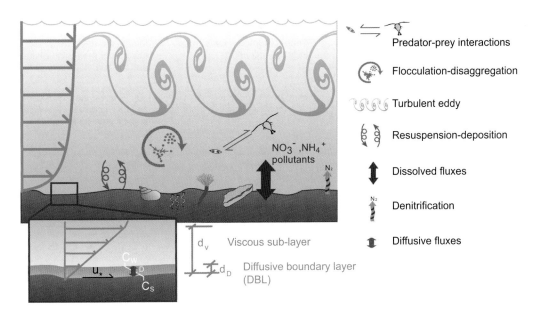

Figure 28: *Mixing and flow strongly influence many important properties and related processes in the benthic and pelagic environments of coastal ecosystems. In general, mixing is more important to planktonic organisms and water column processes, while flow is more important in the benthic environment. C_s is the concentration in sediment porewater and C_w the dissolved nutrient, oxygen or contaminant concentration in the overlying water. See Tables 2 and 3 for definition of terms and relationships.*

There are important turbulent mixing scales to be considered in mesocosm design

In the pelagic environment the most important mixing parameters are T_m, U_{RMS}, ε, l, K_z, and η_k, whereas in the benthic environment, the important parameters are u_*, $\bar{U}$, K_b, and δ_D (Table 2). The Reynolds number Re (in one of its various configurations) reflects the state of turbulence in both environments. The number of parameters seems daunting, and it is often impossible to match all experimental ecosystem turbulence scales to those in nature. Selecting which turbulent mixing scale(s) to use will depend on the experimental system and the questions to be addressed, but mesocosm designers should understand the physical and ecological consequences of their decisions.

Table 2: *This table shows some of the turbulence scales that are important to consider when designing mixing for an experimental ecosystem. (Additional guidance may be found in Sanford 1997.)*

	Parameter	Name	Description	Typical values
	Re	Reynolds number	Ratio of inertial to viscous forces (velocity scale x length scale / viscosity)	10^0 to 10^4 dimensionless, depends on definition
	u_{RMS}	Root mean squared (RMS) turbulence intensity	Characteristic turbulent (fluctuating) velocity in the water column	10^{-3} to 10^{-1} m s^{-1}
	l	Integral length scale	Size of the large eddies	Distance to the nearest boundary, or 10^0 to 10^1 m in the stratified interior
	K_z	Vertical eddy diffusivity	Diffusion coefficient representing enhanced vertical mixing due to turbulence	10^{-6} m^2 s^{-1} in the stratified interior to 10^{-1} m^2 s^{-1} in an energetic bottom boundary layer
	T_m	Mixing time	Time to homogenize a tracer injected at a point	Minutes to days, depending on the size of the region of interest and the turbulent diffusivity
	ε	Turbulent energy dissipation rate	Rate of destruction of turbulent energy by viscosity and shear	10^{-7} W m^{-3} in the stratified interior of the open ocean to 10^{-1} W m^{-3} in the surface layer under breaking waves
	η_k	Kolmogorov microscale	Approximate size of the smallest turbulent eddy	10^{-3} to 10^{-1} m for the range of ε above
	$\bar{U}$	Mean flow speed	Time averaged flow speed	10^{-3} to 10^0 m s^{-1}
	u_*	Shear velocity	$= (\tau/\rho)^{1/2}$, where τ is bottom shear stress and ρ is water density; characteristic velocity scale in boundary layers	10^{-3} to 10^{-1} m s^{-1}
	K_b	Bottom roughness	Effective height of organisms or bedforms on the bottom	10^{-4} to 10^{-1} m
	δ_D	Diffusive boundary layer thickness	Thickness of layer that controls sediment-water fluxes	10^{-3} to 10^{-5} m

(Conversion of ε units: 100 mm^2 s^{-3} = 1 cm^2 s^{-3} = 1 erg g^{-1} s^{-1} = 10^{-1} W m^{-3} = 10^{-4} W kg^{-1} = 10^{-4} m^2 s^{-3}).

Well-established relationships among turbulence parameters can be used to match mesocosms to nature

Fortunately there are some well-established relationships (Table 3) that can be used under many circumstances to relate the various turbulence parameters to each other. This allows characterization of the flow and mixing without measuring every quantity directly, and it limits the number of decisions to be made. For example, if the experimental ecosystem and its natural counterpart mix quickly relative to the ecological time scales of interest, the precise value of T_m is not critical. This in turn implies that the precise value of K_z is not critical. Often, matching either ε or u_{RMS} seems to be most important for pelagic ecosystem experiments.

It is usually not possible to match both because it is very difficult to match integral length scale l, which is a measure of the largest eddies; these are usually much smaller in an experimental system than in nature. Even though ε is a more commonly reported value, it often makes sense to match u_{RMS} instead because more parameters depend on its value. Similar considerations apply to the benthic environment, where u_* is considered to be of primary importance. Simultaneously matching the important parameters in both the pelagic and benthic environments in one experimental system is even more of a challenge, but it is not intractable.

Table 3: Important turbulent mixing relationships for key parameters in experimental coastal ecosystems.

Relationship	Comments	Where important
$u_{RMS} = \sqrt{\frac{1}{3}\left(u'^2 + v'^2 + w'^2\right)}$	Where u', v', w' are the variable velocity components in the x, y and z directions	Pelagic
$\varepsilon \approx \dfrac{u_{RMS}^3}{l}$	Integral length scale, l, is difficult to match between nature and experimental ecosystems	Pelagic
$K_z \approx u_{RMS} \cdot l \approx \varepsilon^{\frac{1}{3}} \cdot l^{\frac{4}{3}}$	Turbulent diffusion controls mixing in water column and is difficult to match to nature because of l	Pelagic
$\eta_k = 2\pi \left(\dfrac{v^3}{\varepsilon}\right)^{\frac{1}{4}}$	v = kinematic viscosity	Pelagic
$T_m = \dfrac{h^2}{2 \cdot K_z}$	Often not as critical to match exactly if well mixed h is total depth or width	Pelagic
$u_* = \sqrt{C_D \cdot \bar{U}}$	C_D is the hydraulic drag coefficient; depends on k_b and $\bar{U}$	Benthic
$\delta_D = \dfrac{10v}{u_*} \cdot \left(\dfrac{D}{v}\right)^{\frac{1}{3}}$	D is the molecular diffusivity (note dependence on u_*)	Benthic

Mixing can have a variety of effects on organisms and ecosystems

The effects of large- and small-scale mixing depend on complex interactions between organism physiology and behavior, nutrient dynamics, and the light environment (Fig. 29, Table 4). On one hand, increased mixing has the potential to increase primary productivity by maintaining cells in the photic zone, by increasing phytoplankton access to benthic nutrients, by decreasing the diffusion gradient around cells, and by increasing copepod excretion rates. On the other hand, increased mixing also has the potential to decrease primary productivity by increasing the turbidity due to sediment resuspension. In addition, although large-scale vertical mixing can replenish nutrient supply to surface waters, it can also mix cells into aphotic waters, disrupting the necessary positive balance between photosynthesis and respiration. Ecosystem productivity and respiration reflect the outcome of these positive and negative effects of mixing. Few empirical studies have been conducted to quantitatively assess ecosystem level responses of plankton communities to the addition of mixing energy.

Table 4: *A summary of empirically determined effects of mixing on phytoplankton, zooplankton, and ecosystem processes from a comprehensive review of the aquatic literature (Petersen et al. 1998).*

Component	Variable	Relationship
Phytoplankton	Settling rate	-
	Chlorophyll *a*	+
	Productivity	+ or 0
	Cell size	+
	Diatom / flagellate	+
	Nutrient uptake	✓
	Timing of bloom	+ or ✓
Zooplankton	Abundance	-
	Variance	-
	Predation rate	✓
	Demographics	✓
Ecosystem	Total biomass	+
	Ecosystem P, R	✓
	Nutrient dynamics	✓

(+) symbol indicates a positive relationship between the variable and turbulence, (–) indicates a negative relationship, (√) indicates the presence of a relationship, (0) indicates no relationship. (P = production; R = respiration)

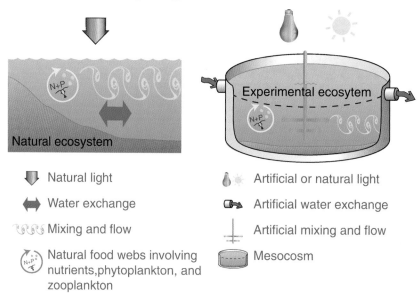

Natural ecosystem
Experimental ecosytem

⬇ Natural light
💡☀ Artificial or natural light

↔ Water exchange
Artificial water exchange

Mixing and flow
Artificial mixing and flow

Natural food webs involving nutrients, phytoplankton, and zooplankton
Mesocosm

Figure 29: *Small-scale mixing around organisms and large-scale vertical mixing between surface and bottom water have a variety of effects on organisms and on ecosystem processes. Generating realistic mixing in mesocosms is challenging, but is crucial to allow for accurate ecological response to treatments applied.*

There are many different ways to mix experimental ecosystems

The two most common techniques that are used to mix pelagic experimental ecosystems are rotating paddles and oscillating grids (or plungers). They each have advantages and disadvantages. Rotating paddles (Fig. 30), if they are carefully designed, have the advantage of promoting both stirring of the whole system and local mixing, providing uniform mixing and exposing all the contained water to turbulent energy. There is less design information about rotating paddles available in the ecological literature, however, especially for the low Reynolds number conditions most relevant for ecosystem studies. Oscillating grids (Fig. 31) are easier to quantify and to implement because their characteristics are well known, but they tend to produce large spatial gradients of turbulent energy without much overall stirring. This may be a problem for many organisms in ecosystem experiments, especially ones whose size approaches the size of the grid or whose physical structure is delicate (e.g., gelatinous zooplankton).

Another method that is sometimes used to mix pelagic experimental ecosystems is air bubbling. There are empirical relationships for turbulence and mixing induced by air bubbling, given the flow rate of air and the size of the system, that make it relatively easy to quantify mixing intensity and mixing time. The major problems with air bubbling are injection of air into the water column resulting in higher dissolved gas concentrations, spray at the surface when rising bubbles break, and potential damage to delicate organisms.

One conceptually attractive type of pelagic experimental ecosystem is the *in situ* enclosure, in which a small volume of a water body is enclosed in a bag and experimentally manipulated. A classic example of this was the CEPEX (Fig. 32) experiment in the late 1970s. However, this approach is fundamentally flawed unless additional mixing is provided because the dominant source of mixing energy (vertical shear) is excluded. Mixing induced by diurnal heating and cooling remains, as well as a small amount of motion transmitted through the flexible walls, but these sources are insufficient to make up for the loss of vertical shear. CEPEX investigators found that their bags were quite under-mixed and that the ecosystems behaved unrealistically as a result.

Figure 30*: Stirrer type mixer used in 1 m³ MEERC pelagic-benthic tanks.*

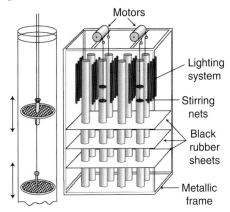

Figure 31*: Oscillating grid mixing scheme (Estrada et al. 1987).*

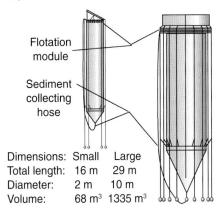

Dimensions:	Small	Large
Total length:	16 m	29 m
Diameter:	2 m	10 m
Volume:	68 m³	1335 m³

Figure 32*: Schematic of* in situ *CEPEX enclosures (Grice and Reeve 1982).*

Experimental benthic systems require special methods for mixing and flow

The most important mixing parameter to replicate in benthic experimental ecosystems is shear velocity, (u_*). Shear velocity can be regulated in flumes in which water flow over the bottom generates turbulence and mixing. An attractive flume design for experimental ecosystem work is an annular flume (Fig. 33), in which a revolving lid drives continuous recirculation of the water. The resulting flow characteristics are not as realistic as those produced with other flume designs. However, because water volume is contained in an annular flume as opposed to a flow-through flume system, it is well suited for experimental ecosystem studies. One disadvantage of laboratory flume systems is their cost, usually limiting replication. In additon, they are typically only practical for studies of the very near-bottom environment.

None of the pelagic mesocosm mixing systems described on the previous page produce realistic bottom shear velocity without over-mixing the water column. As part of MEERC there was an effort to develop a mixing system that generated realistic shear at the bottom while maintaining a pelagic water column with reasonable mixing energy (Porter et al. 2004b). The first attempt at such a device was a modified 1 m³ pelagic tank with realistic water column turbulence coupled to the annular flume.

Figure 33: *Photo of annular flume that was used in benthic-pelagic coupling experiments in MEERC. Flume is 1.80 m O.D, with a 20 cm wide channel and a 15 cm water depth and a 1 m² sediment surface area.*

The linking of the systems proved difficult and, although the device was successful, it was costly and difficult to operate (Porter et al. 2004b).

A system that achieved both goals (i.e., realistic mixing and bottom shear) in a single tank (Fig. 34) was designed and implemented in two sets of three 1 m³ experimental ecosystems studying the effects of sediment resuspension and benthic exchange in eco-toxicology and benthic-pelagic ecosystem studies. This system is simpler to use and is less expensive. Thus, greater replication is achievable without the complexities inherent in linked systems.

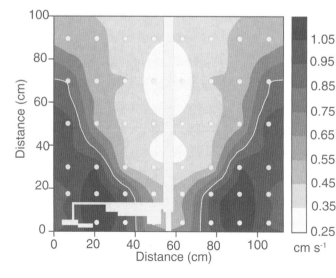

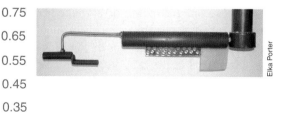

Figure 34: *Contours (left) of turbulence intensity, u_{RMS}, for new mixer design (see photo below) that achieved simultaneously uniform bottom shear stress (or friction velocity, u_*) and realistic water column u_{RMS}. Dots indicate measurement locations.*

Principles from engineering studies are useful guides for designing mixing systems

Engineers have long been concerned with mixing and flow in tanks and channels. In designing mixing reactors for industrial processes and channels (or conduits) for flows of all types of fluids and slurries, engineers have developed many useful techniques, relationships, and rules-of-thumb that are helpful for design of mixing schemes for enclosed experimental ecosystems. The use of scale models (p. 72) is one engineering technique that was employed when developing the MEERC pelagic-benthic mixing systems. Although turbulence in the models did not scale quantitatively to the full scale prototypes, the models provided valuable qualitative information on mixing and flow patterns and paddle arrangements and they allowed relatively inexpensive testing of preliminary designs. Several rules-of-thumb, such as the best ratio of mixing paddle diameter to tank diameter, optimal paddle distance above the bottom or below the surface, and paddle

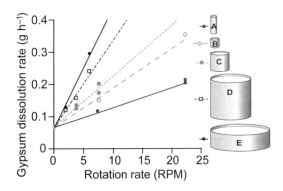

Figure 36: Weighted average rate of gypsum dissolution (symbols) in the five different MEERC pelagic-benthic tanks compared to the cube root of the mixer power as defined by power number theory (lines). A single arbitrary constant for all tanks, which was directly related to the mixer power number, N_p, had to be calibrated to produce the theoretical curves.

spacing, were used (Fig. 35). All are found in the chemical engineering literature (Tatterson 1991).

Many of the engineering studies in which these design principles were developed used much higher mixing energies than appropriate for most experimental ecosystems. However, the design principles still apply because researchers can use dimensional analysis, drawing on principles of similitude. One dimensionless number from the engineering literature that is useful is the Power Number, N_p, which behaves much like a drag coefficient. In the MEERC pelagic-benthic systems, relationships between gypsum dissolution rate (a proxy for u_{RMS} in this case) and mixer RPM in a variety of tanks scaled very well using an N_p relationship from the chemical literature (Fig. 36).

There are many other examples of engineering information that is useful for experimental ecosystem design. Collaboration with civil or chemical engineers, and referring to standard engineering literature, are useful steps when considering the design of a new experimental ecosystem facility.

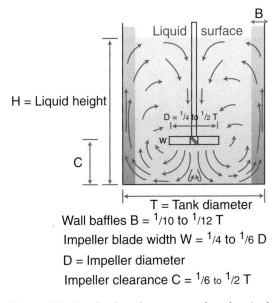

Figure 35: Standard tank geometry for chemical engineering mixing tanks (additional guidance may be found in Sanford 1997) with axial impeller and associated flow pattern.

Mixing and flow must be quantified in experimental ecosystems

Although mixing and flow are clearly important in a qualitative sense, their potential effects on ecosystem structure and function make it essential to quantify their levels and distributions in experimental ecosystems. Intuition about how vigorously or gently to mix a scaled-down piece of nature is not always reliable.

A simple technique to measure flow patterns and mixing times in scale models and full-scale tanks is the use of tracers. In MEERC scale model studies, a food coloring dye diluted to have a density very close to that of the water in the tank was used (Fig. 37). This allowed observations of circulation patterns and tests of the adequacy of different mixer arrangements. A fluorescent dye, Rhodamine–WT (Fig. 38), was used in the full-scale tanks as this dye is detectable at very low concentrations with a properly configured fluorometer. This allowed measurements of the time it took the tank to become thoroughly mixed (T_m) when a pulse input of dye was injected at various locations in the tank. In the MEERC marsh mesocosms, bromide was used as a tracer to measure the flow rate of the groundwater through the sediments.

Tracer studies are useful for measuring flow patterns and mixing times but they are usually not adequate for measuring characteristic flow parameters. Direct flow methods (velocity probes or hot wire/film anemometry) and gypsum dissolution can be used to quantify these other parameters (e.g., Fig. 39); they are discussed in the following pages.

Figure 38: Photo of Rhodamine-WT dye studies in full scale tank. Water samples were collected at timed intervals for fluorometric analysis. Mixing time, T_m, was estimated from these tests.

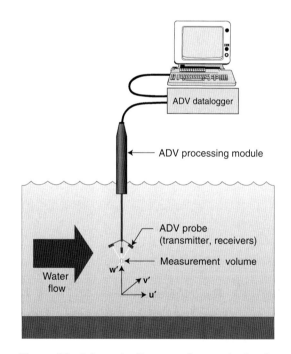

Figure 39: Schematic diagram of acoustic doppler velocimeter (ADV) used for direct flow measurements. (w', v', u' are the variable velocity components in the z, x, and y directions.)

Figure 37: Photo of scale model setup for testing of mixer designs. Diluted food coloring was typically used as tracer in scale models. All tests were video taped.

There are various methods for quantifying mixing and water flow

The best way to quantify turbulence and flow in experimental ecosystems is to measure these factors directly at a number of points on a very small scale with highly accurate instruments. Acoustic Doppler velocimeters (Fig. 39) are relatively inexpensive, easy-to-use instruments for measuring velocity components in three dimensions (x, y, z) in a small sampling volume. From these instantaneous measurements one can calculate the mean flow and turbulent fluctuations directly. Acoustic Doppler velocimeters were used in MEERC to quantify dissipation of turbulent kinetic energy, ε, and turbulence intensity, u_{RMS}, in the water column of the tanks. By taking measurements at strategic locations within the tank, these quantities were mapped out (Fig. 40), averaged over the water volume, and compared to values measured in natural systems. The average values, range of values, and spatial distributions of values were all quantified. Other direct flow measurement techniques are available as well, including laser velocimetry and hot wire/film anemometry, but they tend to be more expensive and more difficult to use.

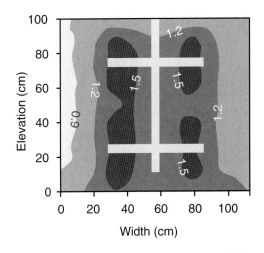

Figure 40: *Contour plot of water column turbulence intensity,* u_{RMS}, *cm s^{-1} calculated from acoustic Doppler velocimeter and gypsum dissolution measurements in 1 m^3 pelagic-benthic tanks with standard stirrer type mixer.*

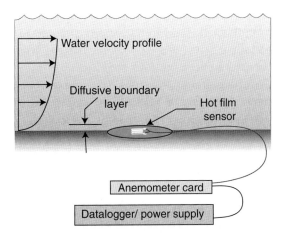

Figure 41: *Schematic diagram of hot film anemometer sensor used for direct measurements of boundary layer friction velocity,* u_*.

For the benthic environment it is important to measure the distribution of shear velocity over the bottom. Hot film anemometers are devices that can be used for this purpose (Fig. 41). The sensor measures the rate of heat loss to the water column, which is controlled by the thickness of the diffusive boundary layer, which, in turn, is inversely related to shear velocity u. These sensors were used extensively in the MEERC program while developing a new mixing device that would better simulate natural benthic boundary layer flows in a standard cylindrical tank (Fig. 34). In addition, these sensors can also be used on the walls of tanks, for instance to help quantify diffusive transport processes from the walls to the water column (Crawford and Sanford 2001).

It is strongly recommended that researchers using experimental ecosystems undertake measurements that allow the direct calculation of the important flow and mixing parameters. In situations where these measurements are not feasible there are other methods available, such as gypsum dissolution. However, these indirect methods can be misleading if not properly interpreted; therefore, they should be used with caution.

Gypsum dissolution can be used in some circumstances

Direct flow measurement techniques (Fig. 42), although they are becoming simpler and less expensive, are still somewhat daunting for researchers without specific training in fluid dynamics. Nevertheless, much of the direct flow data in MEERC was collected by graduate students with undergraduate biology degrees. Gypsum dissolution (Fig. 43) has been suggested as a low-technology alternative to direct flow measurement for ecological studies. However, gypsum dissolution could not be used universally to measure the characteristics of flow in the MEERC tank studies. Nevertheless, if the basic characteristics of a flow are known and gypsum dissolution rates are appropriately calibrated, gypsum dissolution can be a useful tool for interpolation and extrapolation of sparse direct flow measurements.

A workable solution involves limited direct flow measurements to determine whether the velocity field is dominated by unidirectional steady flow or by fluctuations with very little mean velocity, because the response of gypsum dissolution to flow changes significantly under these different circumstances. For example, while gypsum dissolution was linearly proportional to turbulent intensity, the slopes of

these relationships varied with flow regime (Fig. 44). Steady flow produces less dissolution than fluctuating flow for a given turbulence intensity. The gypsum configuration is also important, including the shape of the dissolution objects, their specific chemistry, and the chemistry and temperature of the water (Porter et al. 2000).

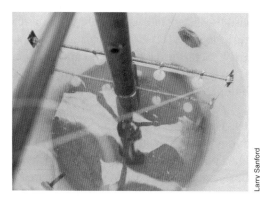

Figure 43: *Gypsum ball arrangement in a MEERC pelagic-benthic tank. Gypsum balls were deployed on metal rods, covering important regions of the tank.*

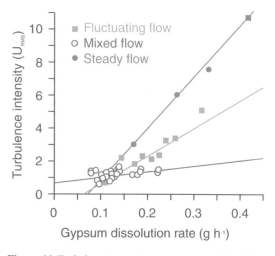

Figure 44: *Turbulence intensity versus gypsum-dissolution of 3-cm spheres in three flow environments; fluctuating environment (n = 15), mixed-flow environment (n = 33), and steady-flow environment (n = 4). Any data points with dissolution rate below 0.07 g h⁻¹ were discarded. The plot reveals that inter-comparison of gypsum dissolution rates for different flow environments is not valid. Similar turbulence intensities can result in significantly different gypsum dissolution (Porter et al. 2000).*

Figure 42: *Mixing design test tank with direct flow measuring sensors; hot-film sensors (yellow circles along tank bottom), and accoustic Doppler velocimeter (3-pronged probe). This setup quantifies water column turbulence and bottom friction velocity for improved mixer design.*

Physical factors: Materials exchange
L.P. Sanford, S.E. Suttles, W.M. Kemp, J.E. Petersen, and L. Murray

Water exchange is an important determinant of ecological dynamics

The rate of water exchange with surrounding ecosystems is a scaling variable that controls key processes in aquatic systems. Indeed, characteristic high rates of primary and secondary production of many estuarine ecosystems have been attributed in part to large material exchange at the interface between watershed and ocean (Odum 1961). Although exchange incorporates both temporal and spatial scale, it is often convenient to express water exchange in terms of residence-time (i.e., time required for incoming water to replace the entire volume of the basin or container), or alternatively as exchange-rate (= residence-time^{-1}).

For a particular pelagic habitat defined by its water volume, the abundance of planktonic organisms changes with the balance between growth (related to resource availability) and mortality (related to predation pressure). The ability of a planktonic population or assemblage to grow and be maintained in this habitat, however, may also be affected by the rate at which these organisms are exported from the habitat with water exchange. If the relative rate of water exchange (flow/volume) between the pelagic habitat and its external environment is greater than the net growth rate of a planktonic population, then that population will be washed from the habitat before it can replace itself. Because different planktonic organisms in the same habitat have different growth rates and swimming speeds, some are more susceptible than others to being washed out. Conversely, in situations where external concentrations of key resources are higher than those within the habitat, water exchange may be valuable for replenishing these resources.

Chemostats are a specific category of experimental planktonic systems where external concentrations of key resources exceed internal concentrations and where plankton growth rates can be controlled by water exchange rate (Fig. 45). In a chemostat, nutrient-dependent growth rate is balanced against exchange-dependent removal rate. Few ecosystem-level studies have attempted to simulate exchange rates in specific natural ecosystems and fewer still have explicitly assessed the effects of exchange rates on ecological dynamics. Indeed, the majority of mesocosm experiments are closed to water exchange. It is essential to consider relationships between biological and physical exchange in the design and operation of experimental aquatic ecosystems.

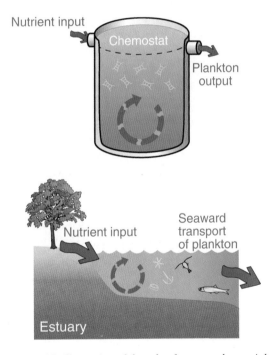

Figure 45*: Illustration of the role of water and material exchange rate in regulating the dynamics of experimental chemostat systems growing phytoplankton and for natural estuarine ecosystems where plankton community production is regulated in part by external inputs of nutrients.*

Residence time should be considered as a controlling scale of experimental ecosystems

Residence time is an important scale to consider in experimental ecosystems because it can determine whether the systems are dominated by physical or biological processes. Many ecosystem processes are best studied in a biological window of space and time scales where biological processes control the variability of biomass, with physical processes exerting more control at both larger and smaller space and time scales (Lewis and Platt 1982). When the space scales of a system are fixed, as in an experimental ecosystem, the residence time of water (T_w) is the primary physical control of the long time scale side of the biological window. T_w is typically thought of as the length of time required for a fluid element to pass through a system, and is normally defined as the volume (V) divided by the flow rate (Q). Its inverse is the exchange rate.

The residence time of a substance or organism in the system (the total residence time T_r) depends on the combination of T_w and the rate of reaction, growth, or death inside the system (r). This rate is commonly expressed as the product of a first order rate constant (k; units of time^{-1}) and the concentration of the substance or organism, as expressed in the mass balance equation (Fig. 46). This means that, for example, the growth rate of phytoplankton is proportional to the concentration of phytoplankton.

The solution to the mass balance equation is plotted in Fig. 47 for $T_w = 10$ (days) and different

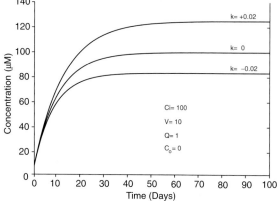

Figure 47: Time-course variations in concentration of a non-conservative property (e.g., nutrients) in a chemostat for a single water exchange rate of 0.1 day^{-1} (Q/V, where Q is water flow and V is container volume), and three first-order reaction rate coefficients. The input concentration is the same in all cases at 100 μM.

values of k, starting from zero internal concentration with an inflowing concentration (C_i) of 100 (μM). The corresponding expression for T_r is also presented in equation form in Fig. 46. T_r is defined as the time needed for the solution to come to within $1/e$ (37%) of its final value. For $k = 0$, $T_r = T_w$ and the final value of the internal concentration is the same as the inflowing concentration. For $k > 0$, growth or production increase both the residence time and the final value of the internal concentration. For $k < 0$, death or consumption decrease the residence time and the final internal concentration. Note that different substances or organisms can have different values of T_r for the same value of T_w.

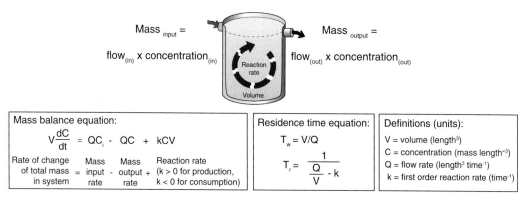

Figure 46: Schematic representation and differential equation describing the rate of change in concentration of a non-conservative property in the water of a chemostat.

Residence time affects plant and animal productivity in natural and experimental ecosystems

In designing experimental ecosystems, the appropriate value of T_w depends on its relationship to the non-conservative biogeochemical rate coefficient, k, and on the research objectives. If T_w is too short ($T_w \ll k^{-1}$), plankton washout occurs and the system is dominated by inputs, such that internal ecological processes are difficult to discern. If T_w is too long ($T_w \gg k^{-1}$), the system approaches a batch reactor which may experience resource depletion with extended experimental duration. If an ecosystem experiment is designed to determine the value of k, a short batch experiment ($Q = 0$) may be appropriate. However, if an experiment is designed to replicate continuous ecosystem function for an extended duration, then T_w must be carefully adjusted.

Results of two sets of MEERC experiments illustrate consequences of alternative choices for T_w. The planktonic ecosystem experiments were run in batch mode (T_w = infinity) and 10% day^{-1} turnover (T_w = 10 day) under relatively high and low nutrient conditions in spring and summer, respectively. Ecosystem productivity increased by 25–35% when exchange increased from 0% to 10% per day under both low and high nutrient conditions (Fig. 48, upper left).

Zooplankton biomass also increased with relatively low external nutrient concentrations, but declined with increased water exchange under high nutrients (Fig. 48, lower left). Under low-nutrient, food-limited conditions, zooplankton varied with primary production. However, zooplankton stocks were inversely related to water exchange at high nutrients due to washout, where zooplankton turnover times exceeded the 10-day water residence time.

Nutrient enrichment experiments were conducted with the submersed aquatic grass, *Potamogeton perfoliatus*, under conditions of low (1 per day) and high (6 per day) water exchange. Increased nutrient input to these experimental ecosystems consistently resulted in higher accumulation of epiphytic algae on the submersed plant leaves, which in turn caused lower growth rates of the host plant, due to shading by the epiphytes (Fig. 48, right panels; Sturgis and Murray 1997; Kemp et al. 2004). Effects of nutrient addition were, however, enhanced at higher water exchange rates, probably because uptake by the high leaf biomass was able to maintain low nutrient levels in the overlying water only at low water exchange rates (Bartleson et al. 2005).

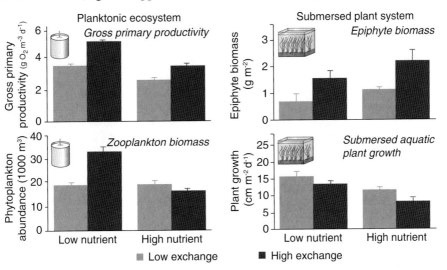

***Figure 48**: Effects of water exchange rate and external nutrient concentration on gross primary productivity and zooplankton biomass in pelagic experimental ecosystems (left panels) and on competition between aquatic grasses and epiphytes growing on plant leaves (right panels). Values presented are experimental means ± standard deviation (Petersen et al. 2003).*

Establishing an experimental exchange rate for water and material requires several practical considerations

Several practical issues affect decisions about water and material exchange rates for mesocosms, in addition to the theoretical questions described earlier. These issues should be considered before the final tank dimensions are fixed because the combination of exchange rate and tank volume (or biomass) determines residence time.

Consumptive water sampling requirements must be considered because the volume of removed water must be replaced. Daily samples of water volume that are too large may result in an excessively short residence time and plankton washout. If sample volume requirements are fixed, then mesocosm volume may need to be adjusted to achieve an appropriate residence time.

The possibility or desirability of separately exchanging water, particles, plankton, and nutrients should be considered. Will the inflowing/outflowing exchange water be unfiltered, filtered but untreated, or filtered and treated? These considerations affect the degree of connection between respective residence times of water, dissolved nutrients, and plankton, as well as the dependence of the experiments on initial conditions or on external exchange conditions. For example, MEERC pelagic-benthic

experiments began with unfiltered estuarine water to establish seed populations, but the exchange water was then filtered to remove any dependence on external biomass variability after the experiment began.

The reliability of exchange mechanisms is critical. Simpler is often better, especially for long experiments. Once-daily (step-function) exchange of water after daily sampling provided a reliable method for pelagic-benthic experiments with little effect on ecosystem function compared to slow, continuous metered exchange. Maintaining slow, metered exchange for a long period is difficult because it involves continuous monitoring and feedback control, but in some cases it is necessary.

Exchange rates must be quantified. For single tanks with controlled external exchange, this is straightforward. However, for an interconnected series of mesocosms or with mesocosms connected to nature by an open pipe or channel, it is much more of a challenge. This is especially true when the systems receive additional mechanical energy, such as in the form of mixing apparatus or wind. Dye or particle tracer techniques are effective for quantifying water flux because they measure water and solute exchange directly (Fig. 49).

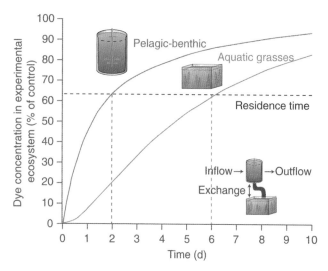

Figure 49: Plot of dye concentration in two connected containers (pelagic-benthic and aquatic grass mesocosms; see inset) used to determine the effective exchange rates and residence times of linked systems. External exchange (inflow) was through the pelagic-benthic system and exchange between systems was controlled by a valve. Continuous dye injection at 100% concentration began in the inflow at time 0. The residence time in each tank was defined as the time to reach 63% of the inflowing concentration. The residence time of the pelagic-benthic tank was 2 days, with an effective exchange rate of 700 L day⁻¹. The residence time of the aquatic grasses tank was 6 days, with an effective exchange rate of 312 L day⁻¹.

Several techniques can be used to effect water and material exchange

Experimental ecosystem exchanges should ideally mimic mechanisms and timing similar to exchanges occurring in nature. There are a variety of methods for accomplishing regular, quantified exchange between an experimental ecosystem and the outside world. Perhaps the simplest of these is pulsed (step function) withdrawal done as part of a routine sampling protocol. For example, in MEERC's pelagic-benthic systems, water and plankton were sampled daily, using a silicon-siphon tube introduced into the center of the tank during active mixing. After sampling, the systems were drawn down by dumping excess water to a sump to achieve 10% volume exchange, followed by refilling with filtered estuarine water.

Another exchange mechanism, used for the very low, continuous exchanges of nutrients and water needed for aquatic grasses experimental

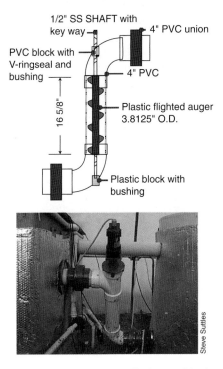

1/2" SS SHAFT with key way

4" PVC union

PVC block with V-ringseal and bushing

4" PVC

16 5/8"

Plastic flighted auger 3.8125" O.D.

Plastic block with bushing

Steve Suttles

Figure 50: *A diagram of a specially designed Archimedes screw pump to facilitate exchange of materials with minimal damage to organisms and aggregates (top), and pictured (bottom).*

ecosystems in MEERC, is metered pumping (e.g., by peristaltic pumps). Here, the exchange itself is an almost imperceptible change in the system. However, the system requires monitoring and feedback to ensure that lines do not become clogged or pumps drift out of calibration.

In both cases, the exchange (or refill) water was taken from a large mixed holding tank, which reduced any tendency for water exchange to contribute to fluctuations in water quality among experimental systems and allowed for input concentrations to be readily quantified. Exchange water was routinely filtered (to 1 μm) before storage in the holding tanks. However, this protocol would have been difficult to control if unfiltered water was used for exchanges.

The plankton in the exchange water should represent the community in the source tank; however, this can be a challenge for organisms with swimming or escape behaviors. In addition, most commercially available pumps use rapidly moving mechanisms with constricted passages or valves that can physically damage delicate organisms. One pump design that avoids this problem is the Archimedes screw pump, used in MEERC's pelagic-benthic-aquatic grass multicosm system (Fig. 50).

Groundwater exchange in MEERC's marsh mesocosms was achieved by a perforated pipe buried below the marsh surface at the upper end of each marsh tank, to mimic the slow infiltration of groundwater into subsurface marsh soils. Tidal exchange was mimicked by computer-controlled periodic inundation and draining of tidal waters at the lower end of each tank (Fig. 51).

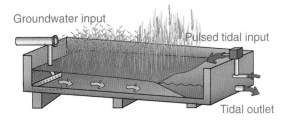

Groundwater input

Pulsed tidal input

Tidal outlet

Figure 51: *Schematic showing water flow and exchange in a MEERC marsh mesocosm.*

Multi-habitat mesocosms with controlled water exchange can simulate spatial gradients and mosaics

Exchanges of water, organisms, and materials among spatially separated habitats and regions define the ecological characteristics of many aquatic ecosystems (Parsons 1982; Zieman et al. 1984; Schindler and Scheuerell 2002). As water, salt, nutrients, organic matter, and other materials are transported between adjacent coastal regions and as mobile organisms move between habitats, spatial gradients and patchy distributions are generated to create the heterogeneous mosaics that characterize most coastal systems. The response of such interconnected multi-habitat ecosystems to external perturbations is likely to differ from that of uncoupled homogeneous environments.

Scientists with MEERC developed a variety of these multi-habitat experimental ecosystems, or multicosms. One such system was designed to simulate realistic biophysical coupling between a planktonic water column and a sediment community dominated by filter-feeding bivalves (Porter et al. 2004a, b). Thus, separate pelagic and benthic mesocosms were engineered to simulate both the internal mixing and the cross-habitat exchanges of water and materials. The small pelagic mesocosm was mixed with radial paddles and its coupled benthic boundary layer mesocosm was a 40 cm diameter Gust microcosm, which

produces uniform shear stress over the bottom through a combination of a spinning disk and central suction (Gust and Mueller 1997). The large pelagic mesocosm was mixed with pulsed water jets and its coupled benthic mesocosm was an annular flume (Fukuda and Lick 1980). The small multicosm was coupled by pumping water from the Gust microcosm into the tank using a centrifugal pump. The large multicosm was coupled using an air-lift pump to pump water from the flume into the tank. Return gravity flow in both systems was controlled by a level sensor. Light intensity in the benthic mesocosms was controlled to equal light intensity at the bottom of the pelagic mesocosms. These multicosm experimental systems were used to examine oysters and water flow on phytoplankton, microphytobenthos, and sediment biogeochemistry, comparing processes in multicosms with those in conventional pelagic-benthic mesocosms. These studies illustrated that experimental systems lacking realistic benthic boundary layer flows and mixing regimes produced markedly different patterns and magnitudes of ecological response to oyster filtration, with benthic algal communities and sediment nutrient cycling showing strongest differences between multicosm and conventional mesocosms (Porter et al. 2004a, b) (Fig. 52).

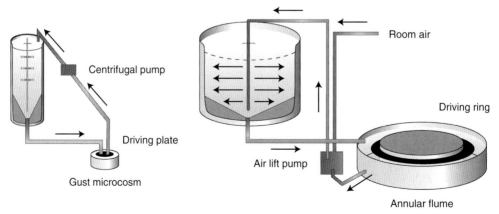

Figure 52: Schematic of two multi-habitat designs with different total volumes and exchange mechanisms. The system on the left coupled a water column simulator stirred with paddles to a benthic-boundary-layer simulator driven by a spinning disk above the sediment surface. Total system volume was 0.1 m³ and exchange was accomplished by a centrifugal pump. The system on the right coupled a water-column simulator stirred by water jetting out of a central shaft to a benthic-boundary-layer simulator driven by a rotating ring. Total system volume was 1 m³ and exchange was accomplished by an air lift pump, which also powered the mixing jets.

Multi-habitat mesocosms with controlled water exchange can simulate spatial gradients and mosaics

In a second example of the multicosm approach, MEERC scientists developed pairs of coupled mesocosms that included shallow habitats with beds of submerged aquatic grasses in one container and combined pelagic-benthic habitat (without vascular plants) in the other (Fig. 53). Field research and numerical modeling studies have suggested that strong interactions associated with tidal exchanges between these habitats alter water quality conditions and productivity of algae and grasses. These interactions were simulated by controlling flow between linked aquatic grass and pelagic mesocosms (Petersen et al. 2003). Retention time in the pelagic component and exchange of water between pelagic and aquatic grass components were each scaled to simulate ranges characteristic of near-shore regions of the Chesapeake Bay. Experiments compared ecosystem dynamics in coupled versus uncoupled systems and for faster versus slower rates of water exchange with external sources.

Preliminary results from these multicosm experiments revealed a number of interesting and consistent patterns. For example, phytoplankton concentrations were lower and dissolved nutrient concentrations were higher in the aquatic grass containers of coupled multicosms compared to uncoupled mesocosms with relatively slow water exchange rates (0.17 per day), while the opposite was true for rapid water exchange (0.5 per day). Abundances of zooplankton were also significantly different for coupled and uncoupled systems. Patterns of epiphyte biomass also appeared to be related to coupling; however, differences were not statistically significant. Although there are still many technical problems to be solved for routine use of a multicosm approach, results suggest the potential for expanded use of modular coupled experimental systems for addressing complex spatial questions in coastal ecology.

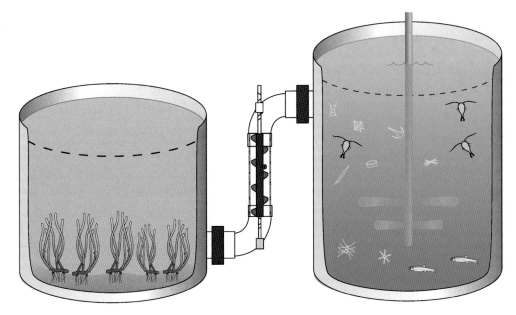

Figure 53: *Schematic of linked aquatic grass and pelagic-benthic mesocosms. The pelagic-benthic mesocosm was stirred by standard axial paddles (shown) while the aquatic grass mesocosm was stirred by a rotating paddlewheel at the water surface (not shown). Exchange between the mesocosms was controlled by a combination of an Archimedes screw pump, a butterfly valve, and an open return flow trough (not shown). External exchanges occurred with the pelagic-benthic mesocosm. Exchange rates were quantified by dye dispersion studies.*

Physical factors: Light
W.M. Kemp, D.C. Hinkle, T. Goertemiller, and J.E. Petersen

Selection and control of light sources is crucial

Light is the fundamental source of energy that drives most primary production in aquatic ecosystems. It is therefore crucial to consider the quantity, quality, and distribution of light in designing experimental coastal ecosystems. Associated with light availability is the photosynthetic production of organic matter to support ecological food-webs and mineral cycles. The ability to establish balanced ecosystems in mesocosms depends in large part on adequate lighting.

There is a range of choices in the design and selection of lighting sources for experimental ecosystems. The design can use either artificial lighting for indoor facilities or natural sunlight for outdoor or greenhouse facilities. Whereas the use of natural lighting tends to ensure adequate quantities of light energy as well as an appropriate spectral quality, it reduces the researcher's ability to control lighting conditions. There is considerable temporal variability in natural light on scales of hours to months as well as spatial variability as a result of different sun angles and shading from nearby structures. Temperature control is also more difficult when natural sunlight is used.

There are numerous alternative artificial lighting sources for indoor mesocosms, including incandescent, fluorescent, metal halide, and mercury vapor lamps. None of these artificial light sources replicates the spectral wavelength distribution of natural sunlight (Fig. 54). For example, incandescent lights are relatively strong between 550 and 650 nm wavelength, but are low above 650 nm and below 500 nm. The spectral distribution of irradiance with fluorescent and mercury vapor bulbs tends to be highly variable. In addition, the quantitative and qualitative outputs from artificial lamps also change with the age of bulbs. For instance, irradiance output from banks of fluorescent lights declines with time, with 15–20% loss of intensity over 2–3 months in MEERC experiments.

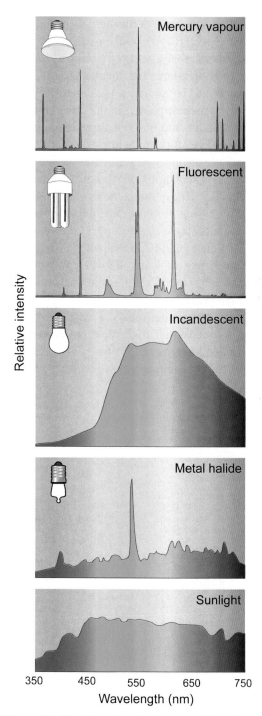

Figure 54: Variations in distribution of the relative intensity of irradiance across the visible light portion of the wavelength spectrum for four alternative artificial sources of light compared to that for natural light.

Spatial and temporal distribution of light needs to be considered

The spatial distribution of photon flux can also be variable at scales of centimeters. For example, photosynthetically available light (PAR) was mapped in indoor MEERC experiments at the water surface of the narrow (0.18 m radius) A tanks and the wider (1.22 m radius) D tanks under light banks with fluorescent bulbs (Fig. 55). Different configurations of light banks and bank heights above the water surface yielded variable light conditions with strong spatial gradients. For example, fluorescent banks suspended 1 m above the water surface produced variations and gradients in light distribution, with narrow tanks (Fig. 55, left panel) having a strong gradient across the water surface, and the wider tanks exhibiting radial distribution from center to outer edges (Fig. 55, right panel).The combination of incandescent and fluorescent bulbs tended to increase irradiance levels and improve spectral quality of the output; however, it also created stronger, more complex, gradients. The increased heat load to mesocosm water associated with artificial lights can be minimized with fluorescent systems by separating the light ballast from the bulbs.

Based on initial experiments, MEERC researchers reduced the spatial variability of light

at the water surface of each tank and among various tanks by adjusting the height of light banks above the water surface and using high-gloss aluminum side-reflectors. This was done in a trial-and-error approach, with physical adjustments followed by a series of light measurements on grids across the water surface of each tank. For any artificial lighting system, light intensity reaching the water declines rapidly with increasing distance between light source and water surface. Practical and safety considerations, however, suggest that the minimum distance between lights and water surface should be at least 50 cm.

Over the course of a day, the intensity of natural sunlight tends to increase progressively until about solar noon, gradually declining toward dusk. In contrast, experiments under artificial lights tend to involve an abrupt step-function (on/off) lighting regime. It is uncertain how significant this problem might be in creating ecological artifacts. It can be partially overcome, however, by using a lighting control system with step-up and step-down sequences that turn individual lights on and off, and combinations of light sources and filters to generate more natural intensities and spectral qualities.

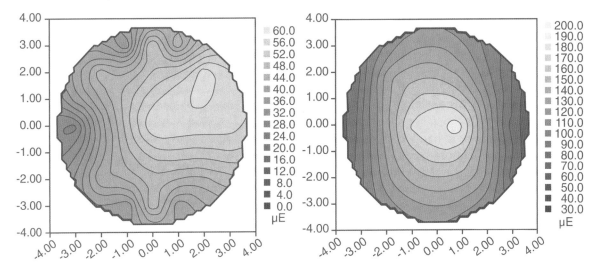

Figure 55: *Distribution of light in a narrow (radius of 0.18 m) A tank (left panel) and in a much larger (radius of 1.22 m) D tank (right panel) (Unpublished data from MEERC pilot studies). Isopleths of light intensity are measured as photosynthetically available light in moles of photons (μmol m^{-2} s^{-1}). Axes (X, Y) indicate relative position in measurement grid across tank width.*

Vertical attenuation of light in mesocosms differs from attenuation in nature because of walls

Light entering an aquatic environment at the air-water surface is scattered and absorbed as it passes through the water column. Dissolved and particulate material and the water itself contribute to the vertical attenuation (absorption plus scattering) of light. In natural waters this diffuse down-welling attenuation follows an exponential relationship in accordance with the Beer-Lambert law by which the fraction of incoming light remaining at any water depth (z) is equal to (exp $(-k_d \cdot z)$), where k_d is the attenuation coefficient.

As with natural aquatic systems, light entering at the air-water surface of experimental ecosystems is also absorbed and scattered by water and its constituents. However, in mesocosms with opaque walls, light may also be reflected and absorbed by the container walls themselves. The effect of walls on diffuse down-welling light is to increase the apparent attenuation and to change the shape of the vertical light distribution from a strict exponential form. Light scattering by walls contributes to the efficiency of light absorption by substances in the water, and pigmented materials on the walls themselves can efficiently absorb sideward traveling photons. Although it is not strictly consistent with the theory, the exponential equation mentioned above can be used to compute an "apparent down-welling light attenuation coefficient", $k_d{}^*$, that includes the combined effects of the water, its constituents, and the surrounding walls.

Walls of the experimental ecosystems used in the MEERC pelagic-benthic studies were constructed from reinforced fiberglass coated with a white glazing material. After the exterior surfaces of these container walls were covered with reflective aluminum coated insulating material, the insulated walls were opaque and textured, with a flat white finish. Mesocosms with clean walls containing filtered (0.2 μm) freshwater, exhibited an increase in the apparent attenuation ($k_d{}^*$) with decreasing mesocosm diameter, particularly below 1 m (Fig. 56). Mean values of apparent attenuation measured during seasonal experiments in replicate mesocosms of the five different widths (increasing from A through E), revealed similar patterns of decreases with container radius, although values of $k_d{}^*$ were generally higher than those observed for respective tanks containing filtered water (Fig. 56; Petersen et al. 1997; Berg et al. 1999; Chen et al. 2000).

The fact that, even when all mesocosms contained identical filtered water, apparent light attenuation $k_d{}^*$ was greater in the narrower containers indicates that the walls themselves contributed significantly to the downward extinction of light. The difference in values for $k_d{}^*$ in filtered water among containers was substantial, ranging from approximately 1.4 m^{-1} for the narrowest (A, radius = 0.18 m) to 0.6 m^{-1} for the widest (E, radius = 1.78 m) containers (Fig. 56). This difference in $k_d{}^*$ between A and E containers means that the percent of surface light transmitted to a water depth of 2 m would be fivefold higher for the wider systems (30%) compared to that for the narrow containers (6%).

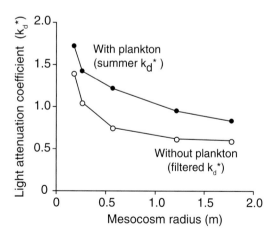

Figure 56: *Decrease in apparent light attenuation coefficient ($k_d{}^*$) with increase in radius of mesocosm containing filtered water (without plankton) or experimental estuarine water (with plankton).*

Light is scattered and absorbed by container walls as it passes through mesocosm water columns

In general, light attenuation by container walls tends to mask any relationships between phytoplankton chlorophyll *a* and k_d*; however, significant relationships were observed in the widest tanks (E) in nutrient enrichment studies. In addition, MEERC researchers also found that k_d* was often correlated with wall periphyton chlorophyll *a* concentrations for the narrowest (A) mesocosms, which showed the strongest wall attenuation on light.

In search of a more detailed explanation for these observed light attenuation patterns, MEERC researchers conducted a series of experiments. To address the question of whether significant light was being transmitted through the tank walls, the translucence of walls were changed for a subsample of mesocosms by removing the reflective insulating material from the external surface. Then the vertical distribution of light in these systems were compared with slightly translucent walls versus that for the conventional systems with opaque walls. Although light levels were higher by 5–15% in containers with translucent walls, the vertical trends in light distribution were virtually identical with and without the reflective external cover (Fig. 57, top panel).

In contrast, if light were being transmitted through walls, it was hypothesized that vertical profiles of light measured near the walls would indicate (at parallel depths) higher light levels than the average light over the container cross-section (Fig. 58). Routine measurements of vertical light attenuation in these mesocosms involved computing mean light levels at successive depths measured at five evenly spaced points across the center of the container's cross-section. In fact, there was significantly more light near (within 10 cm) the walls compared to the average light measured across the container, but only near the water surface (Fig. 57, bottom panel).

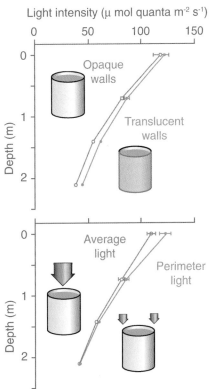

Figure 57: Depth profiles showing differences in light attenuation between mesocosms with opaque walls and those with translucent walls (top panel). Vertical attenuation of light in water columns of translucent C tanks (radius is 0.26 m) where light measurements taken near walls (perimeter light) are compared with average of numerous measurements taken across the tank cross-section (average light) (bottom panel).

Figure 58: View of water column and walls looking downward into a C tank showing walls and mixing apparatus; Secchi disk is also shown for contrast.

Wall color and sheen affect light attenuation characteristics

MEERC scientists conducted another experiment to test the effects of wall color and sheen on apparent downwelling attenuation of PAR. The relative light absorptive characteristics of the mesocosm walls were manipulated by covering the inside of the container surfaces with smooth mylar sheets with both mirror and flat-black finishes. For a selection of narrow (A, radius = 0.18 m) and intermediate width (C, radius = 0.57 m) mesocosms, researchers compared vertical profiles of light levels through filtered water in these mesocosms with normal, highly reflective (mirror) and highly absorptive (flat black) walls (Fig. 59).

These observations showed consistent patterns in which light levels were higher in mesocosms with 'normal' walls than those with black walls, and much higher in systems with mirror walls. The apparent coefficients of light attenuation in mesocosms with the normal white walls were similar to those with black walls. Reflective mirror walls produced highly irregular distributions of light with depth, probably attributable to shadows cast by the mixing apparatus and by the measurement instrument itself. Nevertheless, values of k_d^* computed for C tanks with mirror walls were essentially zero, thus providing further evidence that it is the container walls that produce most of the apparent vertical light attenuation of these mesocosms (Fig. 60).

In summary, these experiments confirm that walls of mesocosms, particularly those less than 1 m radius, have marked effects on light attenuation through the water column. These studies reveal that, even with their flat white coating, walls of MEERC containers have relatively strong light-absorption characteristics. Although investigators often select white as the color of choice for aquatic mesocosm walls (as was done in the MEERC project), few studies have recognized the potentially significant artifact associated with light attenuation by these walls.

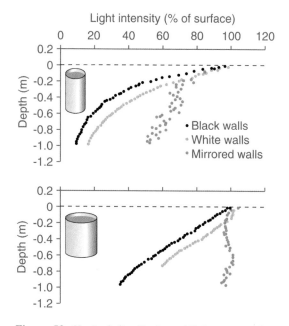

Figure 59: *Vertical distribution of light measured in experimental mesocosms for A tanks (radius of 0.18 m, upper panel) and C tanks (radius of 0.57 m, lower panel), comparing the white wall experimental containers with other containers of the same size but with inside wall surfaces lined with mylar sheets with flat black and mirror finishes.*

Figure 60: *View of series of C tanks showing external reflective insulating material on external surface of container walls and banks of fluorescent lights overhead.*

Physical factors: Walls
W.M. Kemp and C.-C. Chen

Walls create artifacts in the behavior of experimental ecosystems

As discussed in earlier sections of this book, mesocosm walls substantially alter the behavior of experimental ecosystems by limiting water exchange and organism movement, absorbing downwelling light, and by creating physical substrate for luxuriant growth of periphyton communities (Fig. 61). Although periphyton are a normal component in many shallow aquatic systems, the ratio of wall surface to water volume for most experimental ecosystems tends to be much higher than typically found in natural ecosystems. Thus, the unique geometry of mesocosms generally causes the relative contributions of periphyton to total biomass and ecological processes to be greatly exaggerated (Chen et al. 1997, 2000).

Although one might expect that routine removal of periphyton from container walls is a common protocol in mesocosm studies, a recent survey of the literature revealed that only 5% of the 364 published studies indicated that wall-cleaning had been done (Petersen et al. 1999). Although this surprisingly low percentage of mesocosm experiments with routine wall-cleaning may be attributable to logistical constraints or to a misjudging of periphyton effects, it may also be partially the result of concern for how to handle the material removed

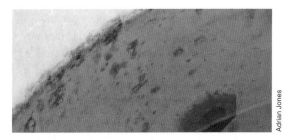
Adrian Jones

Figure 61: Wall growth observed in a MEERC mesocosm.

from the walls. On the one hand, retaining materials in the mesocosm will abruptly deliver algal cells and particulate organic matter to the water column and sediments. On the other hand, removing the material from the mesocosm might cause a substantial loss of nutrients from the system (Chen and Kemp 2004; Dudzik et al. 1979).

These periphyton communities are primarily composed of diatoms, blue-greens and other algae, along with attached bacteria and protozoa. Small animals (epifauna) are often associated with the periphyton, including mobile (e.g., amphipods, isopods, copepods) and attached (e.g., hydroids, bryozoans) animals (Dudzik et al. 1979). In concert, periphyton communities can substantially alter production, respiration, and nutrient cycling processes in experimental ecosystems (Fig. 62; Confer 1972).

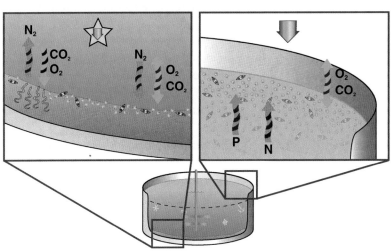

Figure 62: Periphyton attached to container walls creates epibenthic communities that resemble benthic communities in experimental sediments in that they can influence mesocosm primary production and respiration (measured as exchanges of oxygen O_2 and carbon dioxide CO_2), cycling of nitrogen (N) and phosphorus (P), and transformation of biologically available N into gaseous forms (N_2) by denitrification. Periphyton communities, however, do not support large invertebrate infauna or biogeochemical processes that occur deep in experimental sediments (Chen et al. 1997).

Wall cleaning affects nutrient pools and the relative abundance of primary producers

Given the potential importance of experimental artifacts associated with periphyton growth on container walls, it is surprising that few studies have directly examined these effects. MEERC researchers conducted a series of experiments to measure the ecological effects of removing periphyton from interior walls of experimental containers. In all cases, walls were cleaned using an abrasive scrub cloth (3M Scotchbrite Scrubber Pads®) fixed to wooden plates on poles that extended the whole depth of each mesocosm (Fig. 63). The entire wall surface was cleaned with each treatment, and material removed from the walls was retained in the mesocosm and allowed to slough into the water column and sink to the sediment surface. Wall cleaning experiments were conducted in A, B and C mesocosms (respectively, 0.1, 0.1, 1.0 m³ volume, and 0.18, 0.26, 0.57 m radius), with triplicate systems treated with twice-weekly wall cleaning and triplicates maintained as untreated controls. In addition, for the B mesocosms, MEERC researchers compared effects of wall cleaning at once-weekly and twice-weekly frequencies with untreated controls (Chen and Kemp 2004).

In general, these studies revealed that once-weekly wall cleaning reduced algal biomass by 50% whereas twice-weekly wall cleaning resulted in more than 90% decrease in periphyton biomass and primary production. In either case, whereas algal biomass attached to walls declined, phytoplankton and benthic algal biomass tended to increase with wall cleaning. In general, however, total nutrient concentrations in the water column and total ecosystem biomass and primary production were unaffected by wall cleaning.

Although materials removed from container walls were retained in mesocosm water and sediments in MEERC experiments, other researchers have removed wall growth from

Figure 63: Cleaning mesocosm walls using a scrubber pad fixed to a wooden plate and pole.

the experimental system. Indeed, the process of wall cleaning tends to significantly decrease water clarity as materials removed from walls are mixed into the water column. This contributes to an initial increase in phytoplankton and an eventual increase in benthic algal biomass as scraped periphytic material settles to the sediment surface. On the other hand, removal of periphyton from the experimental system represents a potential nutrient sink that could inhibit primary production (Eppley et al. 1978). Based on observed reductions in biomass (and associated nutrient content) of wall periphyton resulting from our wall cleaning protocols, it was calculated that material removed from container walls at once-weekly frequency contains sufficient nitrogen and phosphorus to support nutrient needs of these experimental ecosystems. Thus, although there is clearly an effect on phytoplankton and benthic algal communities resulting from retaining wall scrapings, removing periphytic scrapings from the containers would have too large an effect on the nutrient economy of these experimental ecosystems (Dudzik et al. 1979).

Effects of wall cleaning vary over time and with ambient nutrient conditions

Time-series observations of algal biomass on walls, in the water column, and on sediments revealed that effects of wall cleaning vary markedly with nutrient levels. Although in this experiment significant wall periphyton communities did not develop until nitrogen and phosphorus were added to the experimental systems (Fig. 64), previous studies observed significant increases in wall periphyton biomass, production, and nutrient uptake with daily exchanges (10% container volume) of external estuarine water (Chen et al.1997, 2000). In response to removal of periphyton from container walls, phytoplankton and benthic algal biomass (as chlorophyll *a*) increased markedly in the narrower containers (Fig. 64). For example, the A tanks showed significant

increases in both algal biomass (chlorophyll *a*) and primary productivity (data not shown) for planktonic and benthic communities in systems with wall cleaning treatments. Similar patterns were evident in the B tanks, which are slightly wider but shorter; however, the pattern is confused for the wider and larger volume C tanks (Fig. 64).

Thus, there is a clear tendency for total algal biomass (and productivity) within experimental ecosystems to be unaffected by removal of periphyton from container walls. The partitioning of algae among wall, water, and sediment habitats, however, appears to be altered by wall cleaning protocols. Similar patterns were observed for gross primary productivity (GPP) and nutrient concentrations, where ecosystem level GPP and total nutrient concentrations were unaffected by wall cleaning. The mechanisms by which ecosystem level homeostasis is maintained in the face of large changes in distribution of properties among habitats remains uncertain, although competitive interactions for light and nutrients are likely important (Lewis and Platt 1982), along with direct transfer of algal cells from one habitat to another (Chen and Kemp 2004).

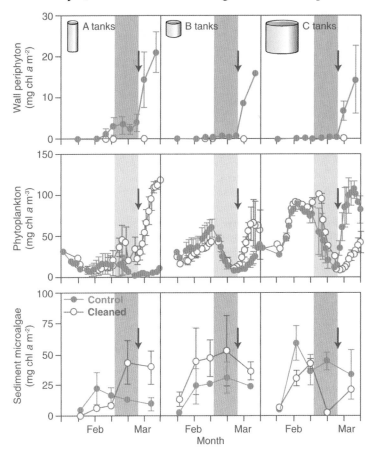

Figure 64: Comparison of time course of biomass (mg chlorophyll a*) of wall periphyton (per m² of wall area), phytoplankton (per m³ of water volume), and sediment microalgae (per m² of sediment area) between controls and cleaned systems in A, B and C tanks (Chen et al.1997). Values are mean ± standard deviation of replicate (*n=3*) mesocosms. Batch mode (no exchange with external water) is indicated by shaded area, at the end of which a pulsed nutrient addition was administered (arrow).*

Effects of wall cleaning are proportional to container radius and cleaning frequency

The relative effect of wall-cleaning both on periphyton and on phytoplankton can be measured as the difference in chlorophyll-*a* (normalized per water volume) between cleaned and control systems ("cleaned" minus "control") for respective habitats. For twice-weekly wall cleaning protocols, this relative effect of periphyton removal was proportional to the radius of the experimental system, with larger effects for narrower containers (Fig. 65). Wall cleaning had progressively less effect on wider experimental ecosystems largely because these systems had lower rates of periphyton accumulation on walls. Although differences in phytoplankton biomass between cleaned containers and controls decreased similarly with container width, no effect of wall cleaning was evident for C tanks with radius of 0.57 m. The fact that slopes of respective relationships for periphyton and phytoplankton are similar but with opposite signs emphasizes again the compensatory effects of wall cleaning (twice-weekly) on the two communities.

Thus, the effect of wall growth, and therefore wall cleaning, on periphyton abundance is inversely related to container radius, and a compensating effect on phytoplankton (i.e., increasing phytoplankton in proportion to decreasing periphyton) also appears to be related to container radius. Whereas partitioning of these ecosystem characteristics among habitats changed with wall cleaning, total algal biomass remained virtually unaffected. Therefore, whereas wall cleaning may be essential for experiments to address questions related to planktonic or benthic communities, it may be less important for experiments focused on ecosystem properties (Chen and Kemp 2004). These relationships (Fig. 65) are similar to those described earlier (p. 56) to emphasize scaling effects of container width on mesocosm processes.

Wall cleaning frequency is an important factor in regulating periphyton growth and its effects on properties of experimental ecosystems. In the experiment described above, MEERC scientists found that periphyton levels were always higher (and quantities of periphyton removed were always larger) with weekly cleaning compared to twice-weekly wall cleaning. The slope of the cleaning effect versus container radius was 3 times higher for the mesocosms treated with twice-weekly cleaning. Thus, for the conditions in this study, twice weekly wall cleaning was required for effective control of periphyton, particularly for narrow (radius ≤ 0.5 m) containers. Under experimental conditions with higher temperature, stronger light, or higher nutrient concentrations, more frequent wall cleaning may be required for effective periphyton control (Kuiper 1981).

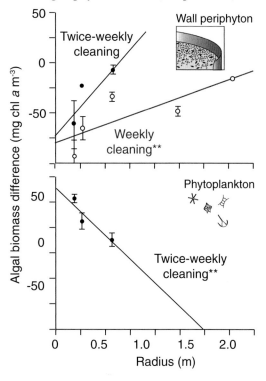

Figure 65: *Linear relations between container radius versus differences (i.e., cleaned – control) in mean biomass of wall periphyton in weekly and twice-weekly treatment systems and phytoplankton in twice-weekly treatment systems.* ** *= significant difference at* $p < 0.01$ *(Chen and Kemp 2004).*

Physical factors: Temperature
L.P. Sanford and S.E. Suttles

Size and shape of experimental ecosystems can affect their temperature

It is well established that temperature is an important variable for many biological and geochemical processes in coastal ecosystems, and therefore must be considered in experimental ecosystems. Experimental systems are small relative to natural systems and thus respond more quickly to heating and cooling, which leads to larger fluctuations in temperature. The same is true when comparing experimental systems of different size and shape, where systems with smaller volumes (higher ratios of wall area- and surface area-to-volume) experience larger temperature fluctuations. As a result it is often necessary to take measures to control temperature in experimental ecosystems.

Figure 66 shows a 9-day temperature-time series for the five different size and shape tanks in the MEERC pelagic-benthic facility. It is evident that the larger volume tanks (D and E) exhibit the smallest daily fluctuations in temperature while the smallest (A and B) have the largest fluctuations. The mean temperature of a tank after the initial adjustment period is determined by numerous factors, including the amount of incident radiation received, the relative importance of surface to wall exchange processes, surface and wall area to volume ratios, whether or not the walls are insulated, and the type of insulation used.

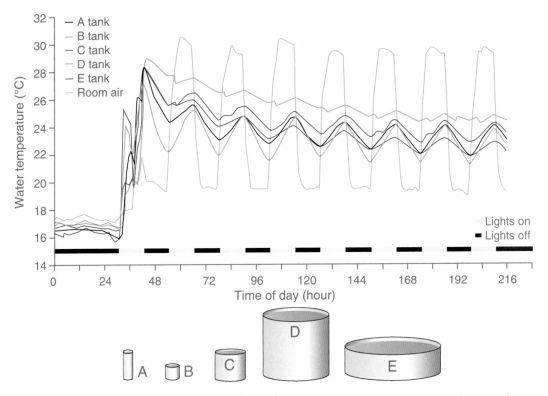

***Figure 66**: Temperature-time series for five MEERC pelagic-benthic tanks of different shapes and sizes and room air temperature for the first 9 days of an experiment.*

Measures should be taken to moderate temperature fluctuations in experimental ecosystems

Researchers constructed a heat budget mathematical model for the MEERC pelagic-benthic systems to help understand the observed temperature differences. The heat budget diagram (Fig. 67) illustrates important exchange processes at the walls and water surface that determine the total heat flux, Q_T, to or from the system. These different processes are defined in the heat flux equation. Although actually measuring or estimating the contribution of each of these processes can be a formidable task, it can provide useful insights for mitigating temperature variations between experimental systems. Researchers estimated the terms in the heat flux equation for the MEERC pelagic-benthic tanks by using theoretical and empirical techniques.

Changes in water temperature in the tank are directly related to changes in the total heat flux, as expressed in the water temperature equation. Using the derived heat flux estimates, researchers solved this temperature equation numerically. They learned that the heat loss during the dark cycle due to longwave radiation, Q_{bwall}, was the dominant controllable heat flux component. This information was used for selecting an appropriate wall insulating material, a reflective bubble wrap insulation. The insulation on the tank walls provided an acceptable solution when all of the tanks were receiving similar levels of incoming radiation.

As another example, MEERC's pelagic-benthic experiments were conducted to examine effects of light levels, including low light levels. It was necessary to add heaters to the low-light tanks to maintain temperatures close to those of the high-light tanks. If these measures had not been taken, the comparison of the results would have been biased by temperature effects, which might have confounded the light effects under investigation.

There are a number of techniques that can be used to moderate temperature fluctuations. These include using wall insulation, adding chillers or heaters, submersing the tanks in circulating water baths, burying them in earthen buffers, or using shading or reflectors (although these also affect light). However, the most important thing is to realize the potential for significant temperature differences and to take adequate steps to avoid or control them.

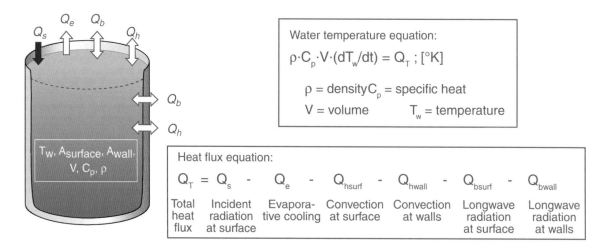

Water temperature equation:

$$\rho \cdot C_p \cdot V \cdot (dT_w/dt) = Q_T \; ; \; [^\circ K]$$

ρ = density $\quad C_p$ = specific heat

V = volume $\quad\quad T_w$ = temperature

Heat flux equation:

$$Q_T = Q_s - Q_e - Q_{hsurf} - Q_{hwall} - Q_{bsurf} - Q_{bwall}$$

Total heat flux	Incident radiation at surface	Evapora-tive cooling	Convection at surface	Convection at walls	Longwave radiation at surface	Longwave radiation at walls

***Figure 67**: Heat budget model for MEERC tanks and governing equations (A = area).*

Biogeochemical factors

J.C. Cornwell and E.T. Porter

There are a number of reasons for incorporating sediments in experimental ecosystems

The presence of sediment in mesocosm studies increases the realism of the experiments, in terms of including more biogeochemical cycling processes and more types of organisms (macrophytes or many benthic organisms obviously require a stable substrate; Fig. 68). Three major reasons to incorporate sediments into experimental ecosystems are described below.

Firstly, sediments have large anoxic zones that are required to reproduce natural cycles of nutrients and contaminants, which are highly sensitive to oxidation-reduction (redox) conditions. For example, reducing conditions may promote the loss of water column nitrate by denitrification, the dehalogenation of organic compounds, and the removal of trace metals by precipitation as sulfides.

Secondly, benthic faunal filter-feeding processes and the reworking of sediment are important to the overall biogeochemistry and ecology of an ecosystem, both in terms of the fate of water column organic matter and the recycling of nutrients and other chemicals.

Thirdly, sediments are almost always the major sink for contaminants, often representing

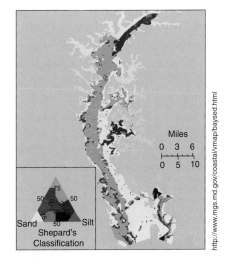

Figure 69: Distribution of sediment grain sizes in Maryland's portion of Chesapeake Bay (adapted from Kerhin et al. 1988).

a legacy of prior pollution. In most cases, toxicological experiments relevant to coastal ecosystems require sediments.

Within a single aquatic system, the size distribution of particles in the bottom sediment can be quite variable, as shown for Chesapeake Bay (Fig. 69).

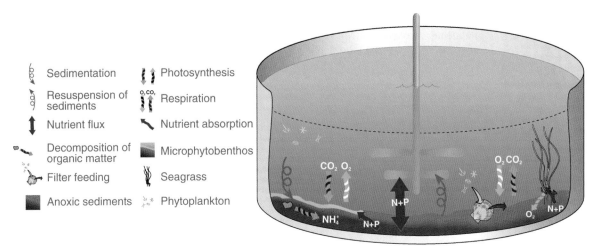

Figure 68: Experimental ecosystems with sediment can simulate biogeochemical processes found in nature.

The first step in designing mesocosms with sediments is to consider the relative benefits of several variables

The nature of sediments required for mesocosm experiments may vary depending on the research questions to be addressed. Sediments may require a particular grain size distribution, or they may need to be subjected to different mixing regimes. For some experiments, resident infauna may need to removed. There are advantages and disadvantages to these actions (Table 5).

The choice of grain size can have effects on sediment resuspension, advection of water through sediment, adsorption/desorption of chemicals, sediment redox, and the suitability for specific benthic organisms. Increasingly, scientists are aware that water, solute, and sometimes fine particulates can flow through sediments of coarser grain size.

The fundamental mineral composition of the system can be variable and this may change both the physical and the chemical structure of the sediments. For experiments involving tropical or subtropical environments, a carbonate type of sediment is generally required. If an experimental system is to simulate deep, anaerobic, estuarine environments (like Chesapeake Bay), it needs a sediment with fine grain size. For high-energy shallow water and beach environments, coarser grained sediments would be generally required.

One should also consider the presence or absence of contaminants in the sediments, as well as marine and freshwater minerals (e.g., pyrite, calcite). In addition, the suitability of temperature, redox, and salinity regimes must be considered in relation to microbial communities. Also, pre-existing labile organic matter can have a profound effect on oxygen and nutrient dynamics during the course of the experiment.

Table 5: *Relative benefits of different sediment mixing regimes, grain size, and maintaining or removing infauna.*

	Sediment treatment choice or grain size	Advantages	Disadvantages
	No treatment, intact cores	Less disruption of microbes, animals, plants, biogeochemical gradients	Inability to prescribe biological community; difficulty in collection and emplacement in large mesocosms
	Defaunation by oxygen depletion	Can prescribe animal species and density	Mild disruption of sediment biogeochemical structure
	Sieving, homogenization	If desired, decreased sediment heterogeneity can prescribe animal species and density	Severe disruption of sediment biogeochemical structure
	Grain size choice or modification	Can use coarser sediments to dilute sediment activity, choose sediment metabolic rates	Mixing fine-grained and coarse sediments can result in heterogeneity after emplacement

Particle grain size is a key influence on biogeochemical processes

Coarse-grain sediment particles tend to have strong resistance to resuspension as well as higher settling velocities. In lacustrine and estuarine physical regimes, resuspension is generally more important for unconsolidated clay and silt particles, although sand may also be transported in very high energy environments or in episodic storm events.

The resistance to advection (flow) of water and particulates through sediments decreases with increasing grain size. In coarse-grained sediments, there is the opportunity for solutes and particles to be moved rapidly into and through the sediments (Fig. 70), depending on physical forcing. Most devices used for sediment incubations do not simulate this process. Small amounts of fine-grained sediment can minimize this advection by filling in the pore spaces between coarse grain particles.

The grain size of sediments is usually a product of the physical regime such as bottom shear, with fine-grained materials often winnowed out during deposition and sediment reworking. In the upper and middle regions of Chesapeake Bay, the organic content of sediments decreases with increasing proportions of coarse-grained materials. This decrease reflects both the difficulty of organic matter to repose in higher energy environments and

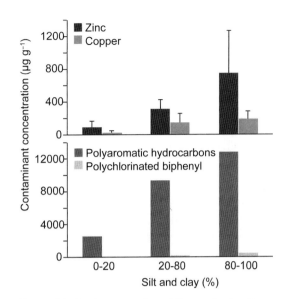

Figure 71: As the proportion of fine-grained sediment increases (from left to right), the concentration of particle-reactive contaminant species increases (Baker et al. 1997).

the low adsorptive ability of low surface area that characterizes coarse sediments.

Contaminants, including inorganic substances such as metals as well as toxic organic compounds, are generally found at high concentrations in fine-grained sediments (Fig. 71). In the heavily contaminated Baltimore Harbor, for example, the highest contaminant levels are observed in fine-grained sediments. The lower organic matter content of coarser sediments generally leads to lower rates of organic matter decomposition, nutrient regeneration, and oxygen uptake.

For mesocosm studies, the choice of grain size can thus have an effect on the experimental outcome. The initial metabolic rate usually is higher in fine-grained sediments, and such sediments can be an unexpected source of nutrients and an unexpected sink for oxygen. The fate of contaminants added to mesocosms may well be affected by sediment grain size, with generally higher sequestration or adsorption of metals and organics in muds than in sands. However, in MEERC STORM experiments, contaminated sediments were added directly to mesocosms and the effect of tidal and episodic resuspension was studied.

Figure 70: Algal cells in the left microscosm have infiltrated the coarse-grained sediment particles as shown by the clear water column. The smaller grain sizes in the remaining microcosm provide greater resistance to such particle advection into the sediments (all three microcosms have defined bottom shear velocities) (Huettel and Rusch 2000).

Metabolic rates of intact sediments vary seasonally at the same site

Although grain size is an important factor, other controlling factors include location within the ecosystem and time of year. Considerations include the following:

- At a given site, there are strong temporal changes in sediment biogeochemistry driven by temperature, changing inputs of organic matter, and changing activity of biota, both animals and plants (Fig. 72).

- In an estuarine setting, the salinity regime of the sediment collection area must be matched with the salinity regime of the influent water. Salinity will affect the microbial, macrobenthic, and floral communities, as well as influence the mineral composition of the sediments and particle aggregation.

- Sediments experience broad ranges of organic matter inputs and quality. The most labile organic matter often is the major contributor to overall sediment metabolism. Addition or depletion of labile organic matter can influence the rates of nutrient exchange. In the Chesapeake mid-Bay, depletion of organic matter (largely from particles sinking in spring and summer) results in a late summer decrease in ammonium flux. Depletion of organic matter during experiments can similarly result in changing rates of biogeochemical reactions. This has a strong influence on microbial processes and solute fluxes.

    In deciding on the nature of the sediments to be used in mesocosms, it is important to consider the effects of season on the experimental outcome. Under warm conditions, organic matter in the sediment will decompose at higher rates than under cooler conditions. A consequence of this is that metabolic rates may change as organic matter is depleted, particularly if the experimental design does not include new sources of organic matter. Under cooler conditions, the pool of labile organic matter is metabolized more slowly, with a less dramatic change in overall metabolic

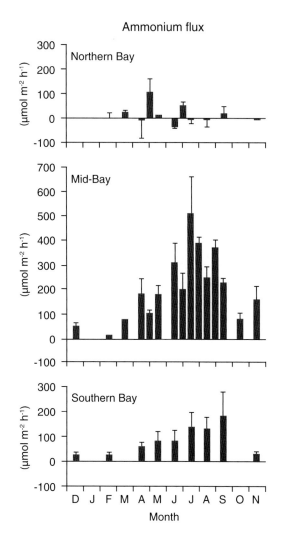

Figure 72: Average values (mean ± standard deviation) of ammonium fluxes at three stations (Northern Bay, Mid-Bay, Southern Bay) along the main axis of Chesapeake Bay. Data were collected between December 1988 and November 1989. Positive and negative nutrient fluxes represent fluxes out of and into sediments, respectively (Cowan and Boynton 1996).

rate over time. Thus, temperature interacts with experimental trajectory, particularly in long-term experiments. The accumulation of solutes such as ammonium and phosphate in pore water tends to increase with warmer temperature, with implications for nutrient releases from manipulated sediments.

Either intact or homogenized sediments can be used in experimental ecosystems

Figure 73: The various steps in using sediments in mesocosm experiments: (a) Mud collection, (b) Homogenization (mud stomping), (c) Mud smoothing, (d) Equilibration.

In designing experimental systems, choosing to use intact or homogenous sediments (Fig. 73) will produce distinctly different sediment traits but will necessarily involve different logistic constraints. This choice is strongly affected by the research questions being asked (Table 6).

Intact sediments preserve vertical zonation within the sediments. One key zonation is the distribution of labile organic matter. This settles to the surface of the sediment and, in the absence of strong bioturbation, makes the surface sediment regime the most biogeochemically active. Intact sediments also preserve realistic three-dimensional biogenic structures within sediments, including animal tubes and burrows. Preserving the original grain size can be difficult, with physical disturbances during collection, transport, and emplacement potentially resulting in the separation of fine-grained and coarser sediment particles. For most indoor mesocosms such as the MEERC pelagic-benthic systems, the

substantial surface area of some mesocosms and the inability to use heavy equipment to transport needed volumes of sediment into the mesocosm room precluded the use of intact sediments.

Grain size can be modified by mixing different proportions of fine-grained and coarse sediments. In preliminary MEERC pelagic-benthic experiments, the initial rate of sediment metabolism had to be minimized to better observe any increases in sediment metabolism and nutrient recycling as a result of organic matter inputs during the experiments. The sediment was constructed from 80% commercial sand with a small amount of fine-grain material from the estuary, using a cement mixer for homogenization. However, when the homogenized sediments were placed in the mesocosms, stirring the water column resulted in substantial movement of the fine-grained sediment, creating pockets of fine-grained material and a very heterogeneous sediment surface.

Table 6: Advantages and disadvantages of using intact or homogenized sediments in experimental ecosystems.

Treatment	Advantages	Disadvantages
Intact	Realistic animal community Realistic vertical zonation & processes Potential for minimal variability	Difficult and expensive to collect and transport Potential high levels of variability Not possible to specify animal or plant community
Homo-genized	Easy and less expensive Can specify animal or plant community	Loss of vertical gradients at onset until about 14 days Re-sorting of grain size Disturbed burrow structures

It is important to understand how sediment homogenization affects experiments

The depth of the sediment layer in mesocosms affects the chemistry of an experimental ecosystem. Very shallow sediment layers will not necessarily provide a good mimic of sedimentary conditions in marshes and submerged aquatic grass beds or of environments with deep-burrowing benthos. Depending on the organisms included in pelagic-benthic systems, depth may or may not be important. The most important depths in terms of nutrient cycles and metabolism are near the sediment surface (Fig. 74). Thus, for most mesocosms, it is not necessary to have deep sediment layers to simulate natural metabolic activity.

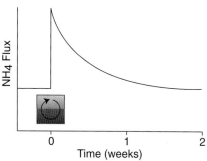

Figure 75: Initial NH_4^+ flux is high after sediment homogenization, with a return to normal after about 2 weeks (Porter et al. 2006).

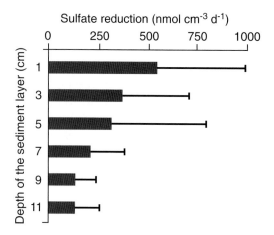

Figure 74: Annual average vertical profiles of sulfate reduction in mid-Chesapeake Bay. Strongly pulsed organic inputs (algae) result in higher rates of sediment metabolism and very high rates at the sediment surface (Marvin-DiPasquale and Capone 1998).

Chemical changes in homogenized sediments are due to rapid diffusion of elevated levels of ammonium from the sediments (Porter et al. 2006). Changes in intact sediments occur more slowly because they are the result of ongoing degradation of organic matter.

Before homogenization, diffusion keeps nutrient concentrations relatively low at the sediment-water interface; homogenization brings high levels of ammonium to the sediment surface and leads to high initial ammonium fluxes (Fig. 75). This effect is not as pronounced for phosphorus

because sediment oxidation during homogenization enhances its adsorption to iron oxides.

In homogenized sediments, diffusion will decrease the concentration of pore-water ammonium at the sediment-water interface, resulting in a decreased flux of species such as ammonium over time. Prior to homogenization the treatment, ammonium production and concentration exhibited distinct vertical profiles (Fig. 76, upper panel), which were lost with initial sediment mixing. After two weeks, the ambient vertical patterns returned for ammonium concentrate but not production (Fig. 76, bottom panel).

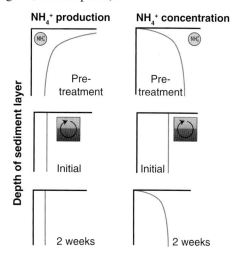

Figure 76: MEERC's ammonium release experiments showed the pattern idealized (Fig. 75). Pore- water ammonium profiles here are shown at the time of homogenization and two weeks later. The deep sediment had higher initial ammonium than the near-surface sediment (Porter et al. 2006).

Sediments can introduce unexpected and unwanted organisms into experimental ecosystems

Sediments do not automatically make experimental mesocosms more realistic. The presence of seeds of aquatic macrophytes and resting stages or cysts of benthic animals in sediments can introduce unwanted organisms (Fig. 77). In some MEERC experiments, the appearance of aquatic grasses and various invertebrates (Fig. 78) was unanticipated. While mechanical filtration of the water column may allow complete removal of pelagic organisms above a particular pore size, and various sieving or sediment preparation procedures can limit the activity of benthic biota, it tends to be very difficult to eliminate sediments as a source of all unwanted organisms if resting stages, cysts, or seeds are very resistant to disturbance.

One possible approach to the question of unexpected biota is to monitor for these species and decide before the experiment whether the experimental protocol includes removal of such biota (if possible).

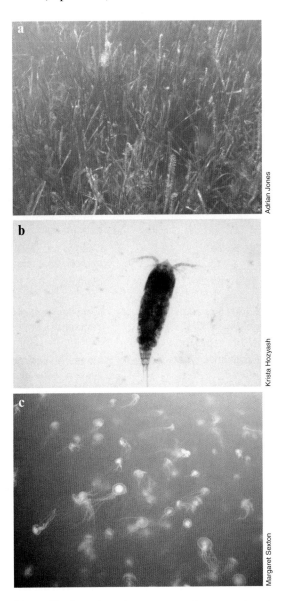

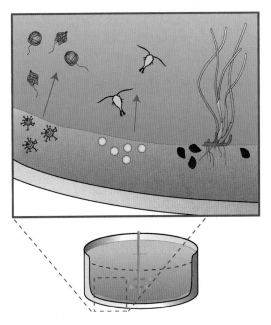

Figure 77: *Seeds, resting stages, or cysts of organisms that occur in sediment brought in from nature and not thoroughly defaunated can unexpectedly change the biotic composition of an experimental ecosystem.*

Figure 78: *Organisms that can contaminate experiments through introduction of seeds, resting stages, or cysts in the sediment include (**a**) aquatic grasses, (**b**) copepods, and (**c**) jellyfish.*

Key sediment measurements are required during experiments

During the course of experiments, the composition and rate of metabolism or nutrient cycling in sediments changes, but the changes occur more slowly than in the water column. The frequency of sampling and the mass or area sampled need to be tailored to the size of the mesocosm, the questions being asked, and potential destruction of experimental habitat.

In addition to container size, research goals may dictate sample amount and frequency. For example, small containers cannot be sampled heavily, especially to provide enough material to measure rates or to allow for replication. Quality assurance and quality control requirements may also dictate the sampling amount and frequency. Table 7 lists the various variables that might be measured in sediments.

Table 7: Potential variables to be measured in sediments of benthic-pelagic coupling experiments.

Variable	Utility	Frequency
Particle grain size	Important to adsorption, water flow	1–2 times per experiment
Water content/bulk density	Needed for conversion of data between volumetric and mass scales	1–2 times per experiment
Solid phase C, N, P	Important for understanding nutrient cycling, organic degradation	1–2 times per experiment
pH profile	Used to calculate solute speciation and physical adsorption	Bi-weekly to monthly
Redox profile	Monitors redox transition depth	Bi-weekly to monthly
Pore-water chemistry (O_2, N, Fe, Mn, S, CH_4, P)	Identifies controls on organic decomposition, solute/gas flux, toxicity	Bi-weekly to monthly
Sediment oxygen demand	Measures sediment component of system respiration	Bi-weekly to monthly
Solute/gas exchange across interface (nutrients, metals)	Measures effect of sediments on overlying water	Bi-weekly to monthly
Sediment rate processes (sulfate reduction, methanogenesis, denitrification, N fixation etc.)	Identifies biogeochemical processes and rates within sediments	Bi-weekly to monthly
Algal photosynthesis	Identifies sediment component of system production	Bi-weekly to monthly
Macrofauna	Identifies the effect of fauna on the ecology and nutrient dynamics	Before and after the experiment
Sediment chlorophyll *a*	Measures microphytobenthos biomass	Weekly
Contaminants	Isotope tracer studies identify contaminant dynamics	Before and after the experiment

Key water column measurements are required during experiments

Chemical measurements are used to monitor changes during experiments so that environmental conditions can be regulated, to measure the rate of change of a chemical (i.e., nutrient or contaminant appearance or disappearance), or to examine the suitability of a chemical habitat for organisms (i.e., pH, O_2, ammonia, salinity).

The suites of chemical measurements are determined by the goals of the experiment. In toxicology experiments, the response of organisms to added chemicals is of interest, so, to ensure that adding a chemical is causing the response, other chemical constituents should be monitored. Experiments on nutrient cycling and the effects of nutrients on plant and animal biomass or productivity or the effect of animals and plants on the ecosystem may have different time courses of change in both chemical and biological constituents (Fig. 79). Thus, chemical measurements may be required at different

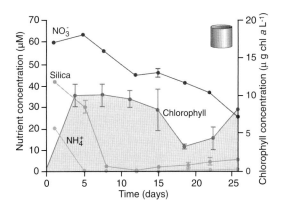

Figure 79: *Representative time courses of changes in NH_4^+, NO_3^-, $Si(OH)_4$, and Chl a concentrations from MEERC C tanks. Error bars represent ± 1 standard deviation of three replicate mesocosms (Berg et al. 1999). These data show algal growth and nutrient uptake or transformation.*

sampling intervals, depending on the process being monitored (Table 8) and if sediment resuspension is incorporated into mescocm experiments.

Table 8: *Potential variables to be sampled in the water column of benthic-pelagic experiments.*

Variables	Utility	Frequency
Temperature	Important control on rate processes, species growth	Continuous to daily
Dissolved O_2	Habitat suitability Production/respiration	Hourly to daily Minutes to hours
NH_3/NH_4	Nutrient uptake or release Ammonia toxicity	Hourly to daily Daily to weekly
Alkalinity/CO_2 system	Production/respiration Carbon limitation	Hourly to daily
Phosphate, nitrate, and silicate	Nutrient cycling	Hourly to bi-weekly
Light attenuation through the water column	Limitation of photosynthesis in the water column and of the sediment	Hourly to weekly
Suspended particulates	Resuspension/sedimentation Biogenic particles	Hourly to weekly Daily to weekly
pH	Chemical speciation Toxicity	Daily to weekly
Salinity/major ions	Osmotic balance, habitat suitability	Daily to weekly
Algal pigments	Plant biomass Community composition	Hourly to bi-weekly Weekly
Contaminants	Inhibition of growth or survival	Weekly
Zooplankton abundance and composition	Algal grazer community	Biweekly to weekly

Physical and ecological complexity
E.D. Houde and J.E. Petersen

Ecological complexity is an important scaling variable that must be considered in experimental design

As a scaling variable, ecological complexity lacks the clearly definable dimensions associated with temporal and spatial scale (i.e., length and time). Nevertheless, the concept of complexity scale has been applied to characterize such variables as species diversity, food-web structure (e.g., trophic depth vs. trophic breadth of a food web), physical complexity of the environment (number of habitats, degree of connection between habitats, diversity of biogeochemical environments, etc.), and levels of ecological organization under consideration (population, community, or ecosystem). Interrelated design decisions must be made regarding each of these subcategories of complexity. A rule of thumb is that the best experimental model contains the minimum degree of complexity necessary to accurately capture the dynamics under investigation. The challenge is obviously to determine a sufficient level of complexity to answer the research question at hand (Fig. 80).

For example, results of MEERC experiments suggest that a relatively simple planktonic experimental ecosystem may be sufficient to capture a range of depth-related scaling patterns evident in deeper natural planktonic ecosystems. But clearly, many important coastal dynamics are determined by more complex interactions within and among trophic levels and among habitats that were excluded from these experiments. Questions involving top carnivores must include additional trophic level complexity. This requires a careful consideration of the particular organisms available for experimentation. Experiments in which fish were added to planktonic ecosystems suggest that in many cases there is a definable window for experimental extent (e.g., size $\geq 10\,m^3$, duration ≤ 30 days) in which fish behavior is not significantly altered.

In experimental ecosystems, habitat complexity generally is greatly reduced. Such systems seldom are large enough to simulate

stratification, water mass fronts, circulation features, or mixing and turbulence over the range of spatial scales that occurs in the sea. Large organisms that normally would migrate or occupy different habitats during various life stages, or on seasonal, diurnal, or tidal time scales, remain confined in the relatively small and homogeneous environment of a mesocosm enclosure.

Figure 80: *Different levels of physical, chemical, and biological complexity (flasks to ecosystem) are necessary to answer different research questions.*

Experimental ecosystems might not accurately reflect natural food web complexity

In aquatic ecosystems, although numbers of higher-level consumers are diminished relative to numbers of organisms at lower trophic levels, their biomass often is high and, in many cases, their size-selective predation controls numbers and biomass of lower trophic levels. Also, organisms at higher trophic levels tend to be large, long-lived, and often highly mobile, making it difficult to enclose them in experimental systems where their behavior may be altered. Thus, some have questioned whether enclosure experiments designed to simulate aquatic ecosystems can really capture the ecological complexity of natural ecosystems (Carpenter 1996).

Complexity is represented by trophic depth and breadth (Fig. 81). Simple experimental ecosystems, with *trophic depth* reduced to only one or two trophic levels, can be used to test effects of environmental factors on a range of ecological dynamics. However, in most natural aquatic ecosystems, predators exert top-down controls over lower trophic levels on time scales beyond those normally included in experimental ecosystems. Enclosure experiments of short duration (days to weeks) that do not include higher level predators may not capture the full range of ecosystem responses or generate stable equilibria that characterize ecosystems in nature.

Trophic breadth is a measure of the number of species present at each trophic level. Species at a given trophic level that are functionally equivalent in nature may respond differently to enclosure volume, depth, and shape and these differences in response may affect the dynamics of lower trophic levels. For example, jellyfishes and fish may be functionally equivalent as predators on plankton, but their very different behaviors under containment can potentially lead to erroneous conclusions as to their role as top-down controllers of ecosystem function or state.

In designing experiments that include trophic complexity, related issues of physical complexity and habitat complexity may need to be considered because these factors may expand in importance as numbers of trophic levels increases.

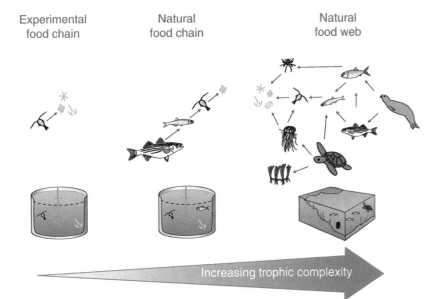

Figure 81: Communities included in experimental ecosystems are typically reduced in trophic complexity relative to natural communities, thus potentially distorting dynamics observed. However, inclusion of higher trophic levels (increased trophic depth) or more species diversity at each trophic level (increased trophic breadth) is not always feasible or desirable. Predators at high trophic levels are often large and may not exhibit normal behavior in small enclosures.

Experimental food chain   Natural food chain   Natural food web

Increasing trophic complexity

Relationships among feeding levels can be isolated and studied in mesocosm experiments if care is taken

It is important for researchers to be aware of the ways in which reduced trophic complexity can potentially distort ecological dynamics observed in mesocosm experiments. Despite the potential for experimental artifacts, mesocom studies can be explicitly designed to test the effect of different levels of trophic complexity on ecological dynamics (Fig. 82). Organisms can be affected from the bottom-up by the resources on which they depend for sustenance (e.g., phytoplankton can be limited by nutrients, grazing zooplankton can be limited by phytoplankton, zooplanktivorous fish can be limited by zooplankton abundance, etc.). Organism abundance and growth can also be controlled from the top-down through grazing

and predation. Bottom-up control can be viewed as positive feedback between organisms and their resources. Top-down control can be seen as negative feedback between one trophic level and the next higher trophic level. A trophic cascade occurs when organisms at high trophic levels increase in abundance, leading to a decrease in abundance of organisms at one trophic level below, which leads in turn to an increase in abundance of the next lower trophic level (Carpenter et al. 1985). Mesocosms provide a potentially valuable tool for exploring important theory related to trophic dynamics in that relationships can be isolated and organisms and nutrients can be selectively added and removed. This section discusses some of the significant challenges as well as opportunities in using mesocosms to study trophic dynamics.

Trophic cascades have important implications for management of aquatic resources, but tests of their potential are notoriously difficult to devise in natural ecosystems. Effects often are diffuse or dispersed, or only detectable at one trophic level removed from a manipulated level at the top or bottom. Manipulation of trophic levels is simpler in enclosed ecosystems and tests of the concept are relatively straightforward. However, it also is easy to introduce artifacts into such experimental tests, especially when large organisms at higher trophic levels are enclosed at unrealistic densities and under conditions where wall effects exercise control over trophic interactions.

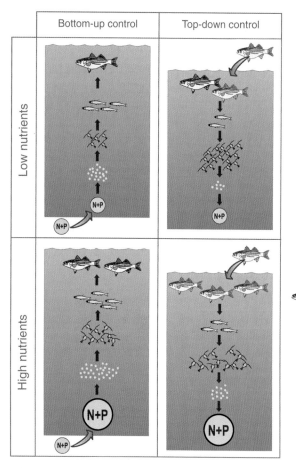

High tropic level predator

Zooplanktivorous fish

Zooplankton

Phytoplankton

N+P Nutrient addition

Predator addition

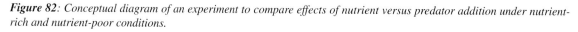

Figure 82: Conceptual diagram of an experiment to compare effects of nutrient versus predator addition under nutrient-rich and nutrient-poor conditions.

There are many practical constraints to introducing predators into experimental ecosystems

There are unique potential problems associated with experimental ecosystem studies involving higher trophic-level animals. For example, it may be difficult to collect desired species and maintain their health, especially if they are large or highly mobile, as are many pelagic fishes. Also, once the organisms are introduced into experimental enclosures, it may be difficult to know if mortalities have occurred unless sophisticated monitoring and tracking methods (e.g., acoustics tags or hydroacoustics assessments) are employed. Such methods may themselves contribute to artifactual animal behavior in enclosures.

Even the largest of experimental enclosures may be miniscule in scale compared to natural ecosystems. In Fig. 83, the volume:perimeter area versus diameter:depth relationships are illustrated for cylindrical ecosystems of 1–10,000 m³. It is immediately apparent that natural ecosystems are off the scale with respect to these relationships and that the MEERC experiments (1–10 m³) are squeezed into the bottom left-hand corner of the figure. The complexity of large, natural ecosystems can hardly be simulated accurately in enclosures of affordable and tractable size. Large pelagic consumer organisms when confined in experimental systems that have low volume-to-wall area ratios tend to have an elevated probability of wall contact that affects feeding behavior (Heath and Houde 2001). Volume:wall area values for all MEERC mesocosms were <1, but natural ecosystems seldom have volume to perimeter values <10–15. Further, diameter:depth values for MEERC mesocosms were <4, but few natural ecosystems would have values <10. To appreciate the scaling distortions between experimental ecosystems and a natural ocean ecosystem with diameter = 1,500 km and mean depth = 4,500 m, the volume:wall area of an idealized cylindrical ocean system is ~188,000 while the diameter:depth value is 333, values far greater than those of enclosed ecosystems.

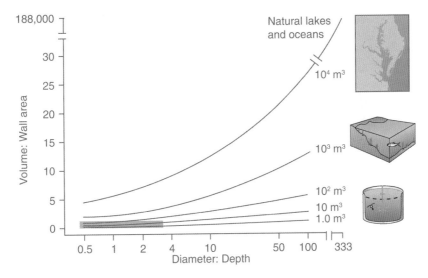

Figure 83*: Changes in the shape (diameter:depth ratio) of experimental and natural ecosystems are associated with changes in the ratio of volume:wall area. Changes in the ratio of volume to area can have important effects on ecological dynamics. The figure illustrates scaling relationships for idealized cylindrical ecosystems. In the figure, volume:wall area for natural lakes and oceans is estimated as the product of average depth and shoreline length. The shaded region on the bottom left of the graph represents the size and shape range of experimental ecosystems employed at the MEERC facility (Heath and Houde 2001).*

A number of other important factors affect experimental ecosystems with multiple levels of biological organization

It may not be possible to construct enclosures that are large enough to conduct meaningful ecosystem-level experiments including very large predators feeding at higher trophic levels without disrupting their behavior. This constraint does not rule out using experimental enclosures to investigate the predation process or to evaluate behavioral mechanisms that shape predator-prey relationships, but it emphasizes the importance of scale in designing research to enclose predators in mesocosms of proper dimensions.

In experimental ecosystems, a useful index of the importance of open water was defined in MEERC research as the ratio of mesocosm volume to wall area. Growth rates of bay anchovy, *Anchoa mitchilli*, and of Atlantic silversides, *Menidia menidia*, were directly related to mesocosm volume. For 50 mm-length anchovy (Fig. 84), growth rates in 10 m³ ecosystems (volume:wall area ratios of 0.61 in D mesocosms and 0.89 in E mesocosms) were

in the range observed in Chesapeake Bay, while slower growth was achieved in 1 m³ mesocosms (volume:wall area = 0.28). Although anchovy growth was in the observed natural range in the more pelagic 10 m³ systems, the rates increased in direct proportion to the volume:wall area ratio and presumably would have been higher yet in a more pelagic experimental ecosystem.

As indicated earlier in this book, the duration of experiments also may affect growth rates. In this example, growth rates of bay anchovy were greatly diminished in the longer experiment (Fig. 84), probably because plankton prey had been reduced to low levels. Although absolute growth rates were diminished, the relative growth rates among mesocosm sizes and shapes were the same, indicating that growth scaled predictably to mesocosm dimensions (Mowitt et al. 2006).

Effects of confinement on predators are apparent even for larval fish only 2–8 mm in length. For naked goby, *Gobiosoma bosc*, larvae growth, mortality, and production all increased rapidly as experimental container volume increased from 20 L to 50 L, demonstrating that an apparent threshold volume of ~50 L was necessary for larvae to attain near-maximum growth rates (Fig. 85).

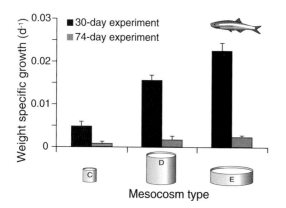

Figure 84: Bay anchovy growth rates in 30-day and 74-day experiments in which fish were added in equal density (number of fish/volume) to cylindrical mesocosms of different sizes (Baker et al. 1997). For this experiment, a definable window for experimental extent (size >10 m³, duration <30 day) in which the experimental ecosystems produced realistic fish behavior was evident. The dimensions of this research window obviously depend on the organism and habitat under investigation, but it appears that the dimensions necessary to achieve realistic behavior can be experimentally determined (Mowitt et al. 2006).

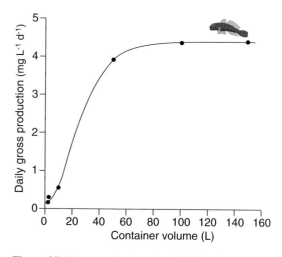

Figure 85: Gross production (mg L⁻¹ d⁻¹) of Gobiosoma bosc *larvae after 20 days in containers ranging from 1 L to 150 L in volume (Heath and Houde 2001).*

Enclosure dimensions affect foraging efficiency of predators

As with all studies, the objectives of enclosure experiments that include higher-level consumers must be clearly defined at the outset. Decisions regarding dimensions of enclosures and the time scale of an experiment should be based on the kind of predator, its size, and the specific objective of the experiment. Even large predators may be enclosed in relatively small mesocosms if the objective is to obtain data on predator behavior and consumption potential in experiments that are of short duration.

Predators in pelagic ecosystems can be classified according to their trophic status, (e.g., planktivores or piscivores). Planktivores, although usually small relative to larger, piscivorous predators, may be subject to strong artifactual scaling effects of enclosure that reduce their swimming and foraging efficiency. The ratio of body size relative to enclosure diameter provides a scaling variable that is inversely proportional to foraging efficiency of planktivores. This is because disruptive contacts

with enclosure walls and the need to avoid walls increase exponentially as enclosure diameters decline for a planktivore of a particular size. Simulation modeling conducted as part of the MEERC program indicated that planktivore lengths should be <5% of enclosure diameters to minimize effects of wall contacts and disruptions (Fig. 86).

Interestingly, modeling indicated that larger, piscivorous predators may increase their foraging efficiency as enclosure dimensions decline (Fig. 86). This outcome occurred because piscivores have potential fish prey in their field of vision for a higher proportion of time in small enclosures than in larger enclosures. Thus, prey encounters are increased and prey becomes more vulnerable to piscivores in smaller enclosures. In its extreme, the situation is analogous to the 'Christians and the lions,' scenario in which extreme predation rates are artifactual and unrepresentative of predation in a natural ecosystem (Heath and Houde 2001).

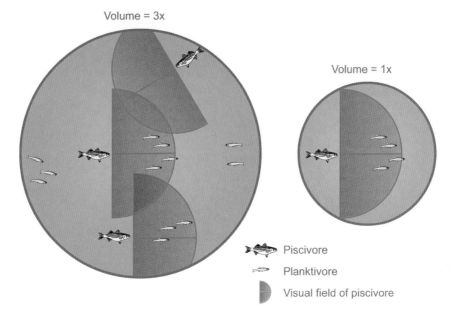

Figure 86: *Enclosure dimensions may modify foraging efficiency of piscivores and their prey (planktivorous fish). Enclosures of increasing diameter and surface area provide better foraging habitat for planktivorous prey species while enclosures with smaller diameters and surface area increase the visibility of the prey and probable encounter rates with piscivores, even when densities of the two predator types are equal (Heath and Houde 2001).*

Time-Space considerations are important in predation experiments

Large consumers, either planktivore or piscivore, are difficult to include in small mesocosms for periods long enough to evaluate their top-down effects on plankton community dynamics. In nature, predators tend to be in contact with prey species for brief periods of time. In contrast, in enclosures, predator and prey may be in close contact all of the time. In a MEERC experiment of ~1-month duration, planktivorous bay anchovy growth and production rates were similar to observations in estuaries. In experiments of ~2-month duration, anchovy depleted the plankton populations, and then their own growth and production declined. Tradeoffs are possible to avoid artifactual outcomes; experimental duration can be adjusted, fish sizes can be selected, and fish densities can be controlled. Sizes of enclosures should be relatively large to minimize artifacts, but they need not have "dimensions of a blimp hangar" (Harte et al. 1980).

Some mesocosm experiments have been explicitly designed to evaluate top-down effects. These effects are generally apparent on the trophic level that is being consumed (e.g., planktivorous fish affecting plankton populations; Fig. 87). However, trophic cascades that extend to prey levels below the level being consumed. (Fig. 88) have not been as clearly demonstrated. Top-down effects were seen in MEERC experiments with anchovy as planktivorous predators.

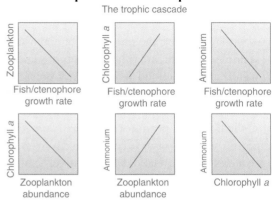

Figure 88: *Predation effects can be measured as top-down control on prey populations, which may generate trophic cascades through lower trophic levels of sequential positive and negative effects, sometimes reaching levels of nutrients. Cascading responses to planktivore predation (top panels) and zooplankton grazing (bottom panels) were seen in MEERC experiments (Muffley 2002).*

Experiments to evaluate effects of higher trophic levels on ecosystem structure can be problematic because in natural ecosystems predators and prey are heterogeneously distributed and may only occasionally interact. One way to address this dilemma is to substitute time for space in pulsed experiments. Adding and removing predators can simulate the intermittent interactions that take place between predators and prey in spatially heterogeneous natural ecosystems. MEERC researchers developed protocols for such experiments, which they termed *fish dipping*, and demonstrated that the protocols were feasible, although logistically challenging.

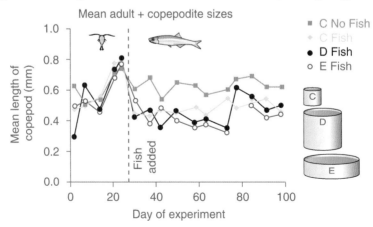

Figure 87: *Effect of fish predation on zooplankton sizes. Mean lengths of copepods in a MEERC mesocosm experiment that included bay anchovy predators (C, D, and E mesocosms) compared to a control (no anchovies, C mesocosm) (Mowitt et al. 2006).*

Small fish and jellyfish are particularly suitable predators for pelagic mesocosm experiments

When selecting predators for enclosed ecosystem experiments, best outcomes may result from use of well-studied, abundant, and representative species. In most cases, small fishes and jellyfishes (Fig. 89) will be the best choices for realistic experimental tests. Fishes are abundant vertebrate predators in freshwater and marine ecosystems, and jellyfishes (ctenophores and medusae) are common invertebrate predators. Both are major consumers of pelagic prey and have the potential to control community structure through top-down effects and trophic cascades.

Selecting predators for experiments requires some initial knowledge of their consumption potential so that appropriate numbers and sizes can be allocated to experiments. Obtaining this information may require preliminary feeding trials or application of bioenergetic models, if they are available.

Once obtained, adding predators to experimental enclosures can be a straightforward task but the subsequent fates of individual predators may not be easy to ascertain. This is especially true in large enclosures and in those with highly turbid or phytoplankton-rich waters where the ability to count predators and identify carcasses is limited by poor visibility. It may not be until the close of an experiment that predators can be collected, measured, and weighed so that growth and production can be estimated. If undetected losses occurred during the experiment, adjustments must then be made in estimates of consumption and predation effects. When losses are detected during an experiment, predators can generally be replaced.

The MEERC experiments used two small fishes (<70 mm length), bay anchovy and Atlantic silversides (*Menidia menidia*), and the lobate ctenophore, *Mnemiopsis leidyi* (also <70 mm length). These three predators are delicate and difficult to handle without injury, but they are representative and abundant pelagic predators on zooplankton in estuaries such as Chesapeake Bay.

In many experiments, predators are stocked at the same low densities in which they occur in natural ecosystems. This approach may preclude sampling during the course of an experiment unless removals are followed by replacement with predators of similar size. In the absence of predator sampling during an experiment, inferences on predation rates and prey selection must be made from trends in prey abundance referenced against controls. Estimates of growth, production, and mortality of predators are made at the end of experiments.

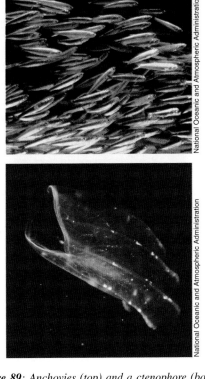

Figure 89: *Anchovies (top) and a ctenophore (bottom). Both organisms can serve as predators in mesocosms; however, selecting predators requires consideration of their consumption potential.*

Special attention to experimental design is necessary to measure trophic interactions successfully in mesocosm experiments

Adequate replication and controls (in which higher trophic levels are absent) are critical to accurately detect effects of predators on community structure and function in experimental ecosystems. In predation experiments, large variability in responses of identical treatment units is not unusual. It may be difficult to evaluate and reach conclusions confidently about effects of predators on ecosystem structure or the scale-dependent effects of containment on predator consumption and production.

In addition to control and replication, it is necessary to repeatedly measure a suite of environmental and biological variables to detect probable trends and cumulative effects of predation and interactions among variables. Nutrient (nitrogen and phosphorous) levels, chlorophyll *a* fluorescence, abundances and sizes of major zooplankton taxa, biomasses,

and environmental variables (e.g., temperature, salinity, dissolved oxygen) may interact to affect levels of predation as well as responses to size and shape of experimental eocsystems, and duration of experiments.

In MEERC experiments with bay anchovy predators, there was weak, but consistent, evidence of a trophic cascade. In this case, copepod population biomass was inversely related to anchovy growth rate (indicator of consumption), chlorophyll *a* level was inversely related to copepod biomass, and chlorophyll *a* level was positively related to anchovy growth rate (Fig. 90). The highly variable relationships with enclosed communities make it surprisingly difficult to test the trophic cascade hypothesis and its possible scale-dependence, even under controlled conditions.

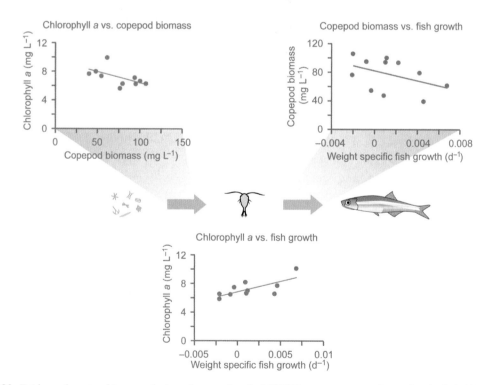

Figure 90: *Evidence for a trophic cascade, based on results of a MEERC mesocosm experiment that included bay anchovy preying on zooplankton, primarily copepods (Mowitt 1999; Mowitt et al. 2006).*

Animals feeding at higher trophic levels generally have behavior and physiology that are particularly scale-sensitive

Large organisms at higher trophic levels may be particularly responsive to scales of temporal and spatial variability in their environments, and these responses are often highly dependent on the particular species involved. In experiments to evaluate production or predator-prey relationships, containment may affect important life processes for these organisms such as respiration, swimming speeds, and schooling behavior, and may distort rates of growth and production.

An understanding of scale and enclosure effects is essential before experimental results can be used for resource management decisions. For example, managers may wish to estimate the consumption and production potential of small forage fishes that feed on plankton and that are themselves preyed upon by larger piscivores. Figure 91 illustrates the effect of declining

mesocosm size on consumption and growth rates of a planktivorous fish, based on calculations from foraging and bioenergetic models. Modest reductions in foraging efficiency in small enclosures result in dramatic reductions in growth rate because energy demands for respiration remain constant as foraging efficiency declines, leaving relatively smaller amounts of energy for growth. For planktivores, the effects of reduced foraging efficiency and declining growth rates in small enclosures also are directly proportional to fish size. Such scale-dependency can have important implications for resource managers who wish to extrapolate from experimental results and predict probable effects of fishes on plankton communities or to determine if forage-fish production is sufficient to meet demands of piscivorous fishes.

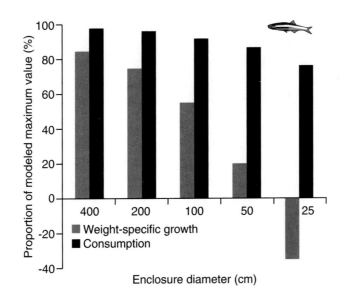

Figure 91: *Computed output of a combined foraging-bioenergetics model (for bay anchovy), showing the reductions in consumption and in weight-specific growth as mesocosm diameter was reduced. Growth declined dramatically under modest declines in consumption, a consequence of constant energy demand for respiration across mesocosm diameters, resulting in lower proportions of consumption available for growth (based on data from Heath and Houde 2001).*

Trophic cascades from zooplankton to nutrients were relatively weak and inconsistent in MEERC experiments

Trophic cascades in which top-down effects of predators cascade through lower trophic levels potentially can have important applications in resource management of estuaries and coastal ecosystems. Due to high variability and interactions with other factors, it can be difficult to isolate and identify trophic cascades in nature. Mesocosm experiments therefore provide a potentially valuable tool for verifying and quantifying the existence of trophic cascades. Increasing abundances of piscivorous fishes, for example, could reduce phytoplankton blooms if there were a strong cascading effect.

Although there was evidence for top-down control in MEERC experiments, the evidence for trophic cascades was weak (Fig. 92). Planktivorous fishes (bay anchovy or Atlantic silversides) and lobate ctenophores (if at high density) did reduce biomass and size structure of major zooplankton groups, but evidence for cascading effects down to the primary producer and nutrient levels was weak or absent. Results of the experiments suggested that both top-down and bottom-up effects were operating. If there were scale-dependent effects on the potential to observe trophic cascades, they were not apparent in the experiments. Thus, the MEERC experiments indicate that complex dynamics can emerge even in controlled predation experiments with simplified natural food webs.

In part, complex dynamics may emerge in enclosure experiments because of the sizes and biomasses of organisms included in the experiments. In enclosures, sizes of predators, in particular, and their biomass with respect to lower trophic levels tend to be high relative to size and biomass spectra in natural ecosystems. In natural ecosystems, high local biomasses of predators are ephemeral while in the enclosed systems persistent high biomasses can alter trophic interactions. Such skewed size and biomass relationships can proliferate during an experiment as organisms grow and the skewed relationships are exacerbated.

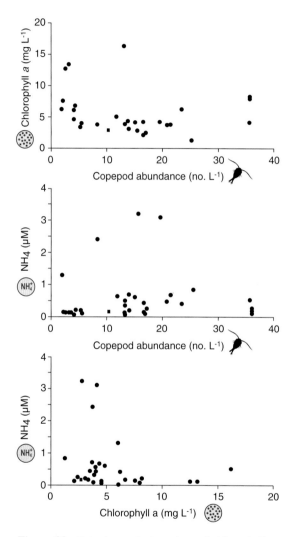

Figure 92: *Paired correlation plots of chlorophyll* a, *copepod abundance, and ammonium (NH$_4$) concentration to illustrate a possible trophic cascade from herbivore to algae to nutrients. In this experiment, there is little evidence of a significant trophic cascade. Fish (Atlantic silversides) growth rates in this experiment (not shown) were directly related to copepod abundances and inversely related to chlorophyll* a, *an effect opposite to that predicted if there were a trophic cascade (Muffley 2002).*

Fish can be bigger consumers than jellyfish on coastal plankton and may be better competitors

Jellyfish are abundant consumers of plankton in many coastal marine ecosystems and have been hypothesized to compete with planktivorous fishes for zooplankton prey. In MEERC experiments, Atlantic silversides proved to be a better competitor than the lobate ctenophore, *Mnemiopsis leidyi*. When enclosed with silversides in mesocosms ranging from 1 to 10 m³ in volume, ctenophores grew slowly and had poor survival (Muffley 2002). Growth rates of ctenophores and silversides were related to the volume/wall area of mesocosms (Fig. 93). Fish and jellyfish grew best in more pelagic 10 m³ D and E mesocosms, an outcome similar to that observed for bay anchovy. Competition among trophic levels was also evident; ctenophores always lost weight in the presence of silversides and virtually none survived, apparently from a combination of nutritional stress and bites from the fish.

The implications of these experiments for natural ecosystems are complex and challenging to interpret. Artifacts of containment clearly reduced growth of both predators and may have exposed ctenophores to, or exacerbated, biting interactions by the fish. Results indicate that growth rates or production levels of these predators, based on experiments in small laboratory systems, are likely to underestimate potentials in natural ecosystems. A similar conclusion was reached for bay anchovy in other MEERC experiments (Muffley 2002).

A further complication in comparing jellyfish and fish predators is the watery body composition of the jellyfish. In MEERC experiments, ctenophores and silversides of 40–60 mm length were used. Consumption potential of an individual fish was estimated to be ~3–4 times greater than that of an individual ctenophore. Consequently, in the MEERC experiments, four ctenophores were considered to be functionally (ecologically) equivalent to one fish. However, there are other possible ways to establish equivalency. For example, physiological or energetics equivalencies might standardize individuals to their carbon content. By that standard, the ctenophore would clearly be a bigger consumer than the fish.

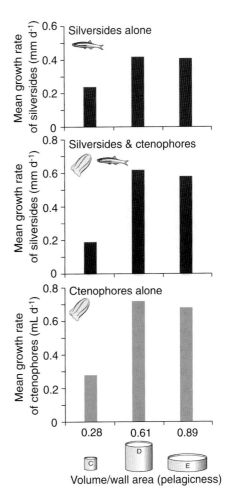

Figure 93: *Mean growth rates of Atlantic silversides (when included with and without lobate ctenophores), and mean growth rates of the lobate ctenophore relative to mesocosm size and shape (C, D, and E mesocosms) in a MEERC experiment. No growth data for ctenophores in the presence of Atlantic silversides are presented because there was virtually no survival, with a loss of weight occuring in the few surviving ctenophores in these trials, indicating that the ctenophore predators were poor competitors (Muffley 2002).*

Experiments can be altered by unintentional introduction of larvae and differences in initial conditions among replicates

To foster realism, trophic dynamic experiments at MEERC were initiated with coarsely filtered water from the Chesapeake Bay containing natural plankton communities. A disadvantage of using natural planktonic communities is the possible unintended inclusion of larval forms of top predators. In MEERC, the small medusa *Moerisia lyonsi* was inadvertently established in several experiments, sometimes at densities sufficient to cause significant predation on zooplankton (Fig. 94; Purcell et al. 1999; Steele and Gamble 1982). At other times, the large lobate ctenophore *Mnemiopsis leidyi* occurred inadvertently. Ctenophore densities were usually low, but sometimes at abundances high enough to cause significant predation on zooplankton. If the objective of an enclosure experiment is to evaluate effects of introduced predators, then these sorts of unintentional introductions can be confounding, even if the inadvertent predators are accounted for in analysis.

Unintended introductions of species have been observed in many enclosed ecosystem experiments. For example, inadvertent introduction of high densities of the ctenophore *Bolinopsis sp.* in Loch Ewe mesocosms were reported to control the trajectory of plankton communities in experiments (Steele and Gamble 1982). In the CEPEX experiments on Saanich Inlet in British Columbia (Reeve et al. 1982), there were confounding effects from the variable and uncontrolled introductions of carnivorous ctenophores *Bolinopsis infundibulum* and *Pleurobrachia bachei*. These predatory gelatinous zooplankton changed the trajectories of plankton communities in the large CEPEX enclosure experiments.

It can be argued that initial populations sometimes simulate blooms of organisms that occur in natural ecosystems and that much can be learned from such occurrences with respect to trophic relationships and behavior of pelagic ecosystems. Although this may be true, under most circumstances in long and expensive enclosure experiments, such phenomena will introduce additional variability that is undesirable for efficient analysis of experimental treatments.

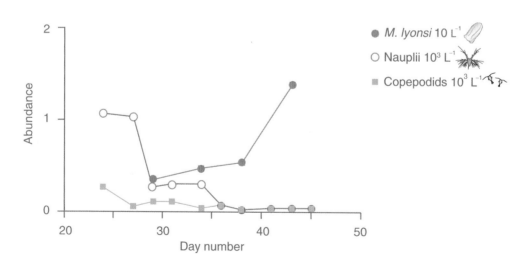

Figure 94: *Abundances of the medusa* Moerisia lyonsi, *and of copepodids (juvenile copepods) in a 1 m^3 C mesocosm from a MEERC experiment. The medusa was inadvertently established and became abundant in the mesocosm. Its predation is the apparent cause of the collapse of the copepodid populations (Purcell et al. 1999).*

Biological factors: Aquatic grasses
L. Murray and W.M. Kemp

Special considerations are needed in designing aquatic grass mesocosms: container size and location

Experimental ecosystems designs containing submerged aquatic grasses require unique criteria that differ from those considered for pelagic-benthic mesocosms. For example, spectral quality of lighting becomes more important because of its effect on hormonal control of flowering and cell elongation in vascular plants. In addition, aquatic grasses need 10–20 times more light for survival than do phytoplankton and benthic algae. Thus, the minimum required light intensities for experimental systems with aquatic grasses are relatively high ($\geq$ 15–25% of natural incident light; (Kemp et al. 2004). Because aquatic grasses obtain much of their nutrient requirement by root uptake, design for aquatic grass mesocosms must consider volume and nutrient content of sediments needed to support plant growth. On the other hand, efficient uptake of CO_2 and nutrients via plant leaves requires vigorous mixing to facilitate transport across leaf boundary layers (Murray et al. 2000). Although aquatic grasses provide habitat for diverse and abundant communities of animals, stable herbivore populations are needed to control algal overgrowth while minimizing consumption of plant leaves. These factors involve spatial and temporal scales that must be considered in the design of mesocosms containing aquatic grasses.

MEERC researchers have designed and used diverse aquatic grass mesocosms, ranging in scale from indoor aquaria (0.1 m^3), to larger tanks (1 m^3) held in a greenhouse, to flow-through outdoor ponds (500 m^3) (Table 9). In general, there are tradeoffs between greater control for smaller systems and greater realism in the larger systems (Table 9). Indoor aquaria allow for excellent control of light and temperature and ease of sampling. They are relatively inexpensive to replicate, but their artificial lighting is poor quality and space limitations reduce food web complexity. Outdoor pond systems have sufficient size to contain larger animals, greater diversity, and more complex webs with fish; however, internal patchiness and poor mixing makes them more difficult to sample, and they are expensive to build and difficult to maintain. Greenhouse tanks tend to have intermediate control and realism, with natural light but larger volume that increases sampling effort and reduces replicability.

Table 9: Summary of tradeoffs in choosing among alternative experimental ecosystems for studies involving submersed vascular plants in coastal environments.

		Advantages	Disadvantages
John Melton	Laboratory aquaria (0.1 m^3)	Control light and temperature Control biology and chemistry Ease of samplling Better replication	Artificial lights with lower intensity and poor spectral quality Simple food webs No large animals
John Melton	Greenhouse tanks (1 m^3)	Fair control of temperature Semi-natural lighting Modest control of biology Some internal variability	Difficult to control wall growth More difficult to sample More difficult to replicate Expensive to maintain
Tracey Saxby	Outdoor ponds (500 m^3)	More realistic biology Complex food webs with fish Natural lighting Spatial heterogeneity	Variable with poor replication Difficult to sample Difficult to mix water column Expensive to build

Special considerations are needed in designing aquatic grass mesocosms: plant biology

In any experimental ecosystem containing submerged aquatic grasses, plants must be established by one or more methods, including transplanting vertical shoots, rhizome pieces, or whole plants, or sowing viable seeds or underground vegetative propagules. Knowledge of physiology and reproductive biology of experimental plants is needed for effective planting. Once plant communities are established, their density can be controlled for uniformity among replicates by selective weeding and trimming.

Because aquatic grasses are characterized by a range of morphologies and growth forms, species composition must be considered in mesocosm design. Growth patterns for aquatic grasses can be separated into distinct groups with basal or apical meristems (Fig. 95). Basal meristem plants such as eelgrass (*Zostera marina*) grow with cell division and formation of new tissue occurring primarily at the base of the leaf, usually near the sediment surface, and they form broad meadows of relatively uniform height (e.g., 0.4–0.6 m). For apical meristem plants, such as redhead grass (*Potamogeton perfoliatus*), new growth occurs primarily near the uppermost tip of the vertical stems and at the growing tip of the belowground rhizome. Apical meristem plants tend to be tall (0.6–1.5 m), often dropping their lower leaves to form canopies near the water surface.

Most aquatic grass species are capable of obtaining nutrients either from sediments via root uptake or from the overlying water column through their leaves. The relative importance of the two pathways depends in part on relative nutrient concentrations in sediments versus the water. Thus, plant growth and nutrient dynamics will be regulated by the ratio of sediment volume to water volume. Moreover, the relative size of sediment nutrient pools may set the maximum duration for aquatic grass mesocosm studies.

In summary, experimental ecosystems need to be designed to accommodate the biology and physiology of experimental plants. For example, taller containers are generally required for apical meristem plants. Because submersed plants have high light requirements for survival, mesocosm height and water clarity are important constraints in mesocosm design. Sediments need to be sufficiently deep and rich to provide nutrient pools that will sustain plant growth throughout the experiment. Thus, scaling ratios of water depth to sediment depth need to be established for growth of particular submersed plant species in experimental ecosystems. In all cases, these design considerations involve matching mesocosm size to the experimental organisms and communities being studied.

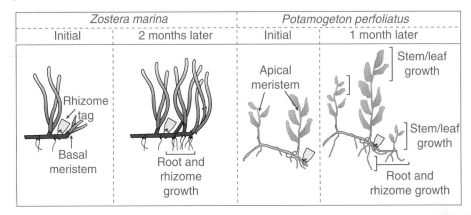

Figure 95: *Schematic diagram illustrating the different distribution of leaves and roots of relatively different age for basal meristem plants such as* Z. marina *(left) and apical meristem plants such as* P. perfoliatus *(right).*

Duration of experiments needs to balance between perturbation-response time and resource-depletion time

Submersed plants are sensitive to their nutrient regime. Whereas growth of many aquatic grass populations in nature is often limited by insufficient availability of sediment nutrients, plant growth may also be indirectly inhibited by high nutrient levels in the water column. In this case, high nutrient concentrations stimulate accumulation of planktonic and epiphytic algae that shade submersed plants, limiting light. Many aspects of nutrient effects on growth of aquatic grasses have been studied in MEERC mesocosms and other experimental systems (Kemp et al. 1983; Short et al. 1995; Neckles et al. 1993). These studies illustrate important considerations for the design and conduct of research involving experimental ecosystems with submersed vascular plants.

Mesocosm experiments compared time-scales of plant and epiphyte responses to nutrient addition. Although earlier experiments demonstrated that nutrient additions to overlying water can inhibit submersed plant growth by promoting accumulation of epiphytic algae on plant leaves (Twilley et al. 1985; Neundorfer and Kemp 1993), time-scales of responses to different nutrient amendment regimes were only recently considered. The basic designs of these experiments were all similar (Fig. 96), and most reported that nutrient-induced increases

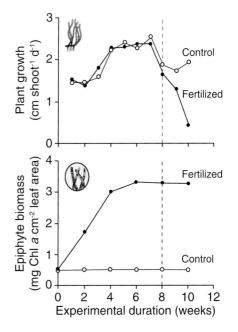

Figure 97: *Time-course response of submersed plants (*P. perfoliatus*) and epiphytes on plant leaves following initiation of continuous nutrient enrichment at week 0 (Sturgis and Murray 1997).*

in epiphytic biomass impeded light penetration and nutrient diffusion through this material to a point where vascular plant growth declined. This effect, however, is not instantaneous.

Plant growth responses in MEERC experiments lagged 6–7 weeks behind initial increases in epiphyte accumulation (Fig. 97; Twilley et al. 1985). Even after epiphyte biomass accumulation saturated in week 6, another 1–2 weeks were required before submersed plant growth consistently decreased in fertilized mesocosms. Presumably, plant leaves can continue to grow for a time (even under low light levels) until shading and diffusion blockage reach thresholds. Plant growth in this experiment began to decline after 7 weeks even in the control mesocosms, perhaps due to depletion of sediment nutrient pools. These trends reveal that experiments must be long enough to allow for observable responses but short enough to avoid nutrient stress. In MEERC studies with *P. perfoliatus*, a duration of 8 weeks was adequate.

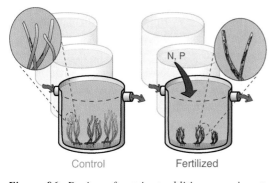

Figure 96: *Design of nutrient addition experiments in aquatic grass mesocosms, where nitrogen (N) and phosphorus (P) were added continuously to water columns of treated experimental systems, with plant and algal growth monitored regularly in control and treated systems.*

Water exchange rates affect submersed plant community responses to fertilization

The rate at which water is exchanged between an experimental (or natural) ecosystem and its external environment measures the system's degree of isolation. The character of an isolated system can change markedly in response to perturbations. Whereas many coastal environments tend to be driven by relatively strong water circulation, in most experimental coastal ecosystems water exchange is limited by walls. The wide disparity in water exchange rates in various experimental coastal ecosystem studies (0–16 water-volume exchanges per day) may partly account for substantial differences in reported submersed plant community responses to nutrient loading (Murray et al. 2000).

Under conditions where nutrient enriched external water exchanges relatively slowly with a dense submersed plant bed, nutrient uptake by the plants can suppress epiphyte growth by reducing nutrient availability. Conversely, at very high water exchange rates, leaf uptake would be unable to control nutrient concentrations, thereby allowing epiphytes to grow. For a given nutrient loading rate, the fraction of incoming nutrients assimilated by aquatic grass leaves would be inversely related to water exchange rate and directly proportional to aquatic grass biomass. (Bartleson et al. 2005).

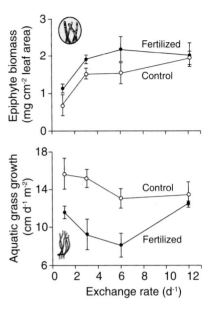

Figure 99: Experimental results showing how rate of mesocosm water volume exchange affects responses of epiphyte biomass (upper) and submersed plant (P. perfoliatus) growth rate (lower) to nutrient enrichment treatment (fertilization).

MEERC scientists tested these hypotheses by conducting experiments (Fig. 98) in 1.0 m³ mesocosms to examine how nutrient-enrichment responses of aquatic grasses, epiphytes, and other community components are affected by water exchange rates (Bartleson et al. 2005). Four water exchange rates were used (1, 3, 6, and 12 times per day) with two levels of external nutrient concentration (3 and 22 µM DIN, 0.3 and 2.2 µM DIP, respectively). Epiphyte growth in control and fertilized treatments increased with water exchange rate up to 6 per day, but effects were greater in high-nutrient mesocosms (Fig. 99). Plant growth decreased in correspondence with increased epiphytic growth. At the highest experimental water exchange rate (12 times per day), epiphyte biomass and aquatic grass growth were unaffected by nutrient enrichment. Presumably, nutrient loading was sufficiently high at this water exchange rate so as to saturate nutrient demands for epiphyte growth, which became limited by other factors such as light.

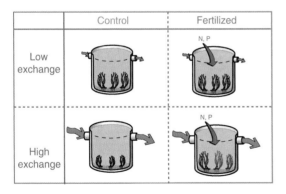

Figure 98: Simplified design of mesocosm experiments to test how submersed plant community responses to nutrient enrichment (N, P) are affected by the rate at which experimental water-volume is exchanged with external sources.

Grazers influence submersed plant community responses to fertilization but are sometimes difficult to control

In natural aquatic grass beds, epibenthic herbivorous invertebrates like snails and amphipods consume epiphytic algae on plant leaves. Populations of these small invertebrates are generally kept in check by fish predation. Climatic factors or anthropogenic changes in abundance of larger animals that prey on these fish can unbalance these food webs. Hypothetically, this could affect the response of epiphytes and aquatic grasses to nutrient enrichment. MEERC scientists examined this hypothesis by adding gammarid amphipods (*Leptocheirus* sp.) to half of a series of mesocosms treated with different levels of nutrient additions (Fig. 100).

Epiphytes accumulated with nutrient enrichment in mesocosms; however, levels were significantly lower in the presence of the grazers (Fig. 101). Aquatic grass growth was generally inversely related to epiphyte accumulation, so that amphipod grazing enhanced plant growth. Grazing effects on epiphyte levels were most pronounced at intermediate nutrient enrichment levels. Apparently, at low nutrient levels, epiphytic growth was too low to support much amphipod

activity, muting grazing effects. At highest nutrient levels, however, epiphytic growth rates were enhanced and overwhelmed the grazers' ability to control epiphyte production.

In nature, diverse animal populations often inhabit aquatic grass beds. These animal populations can, however, be difficult to maintain in experimental ecosystems. For example, it is common when establishing aquatic grass mesocosms to find a few individual gammarid amphipods introduced inadvertently. Although a few animals might have little effect on experimental ecosystem processes, if they reproduce rapidly there can be explosive population growth within weeks. Whereas these herbivorous organisms generally feed on epiphytic and benthic algae in nature, they will switch to feeding on aquatic grass leaves if the algal food stocks are depleted in mesocosms. In small mesocosms it is difficult to maintain predatory fish to regulate grazers because the confined space induces aberrant behavior that can disrupt experimental ecosystem dynamics. Hence, special care must be taken to control herbivore populations in experimental submersed grass systems.

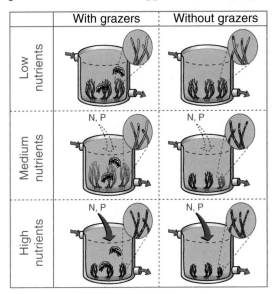

Figure 100: Response of epiphyte biomass levels to interacting nutrient and grazer treatments.

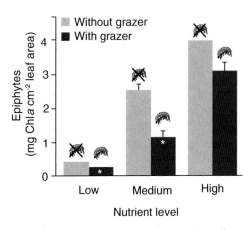

Figure 101: Experiments to test interacting effects of amphipod grazing at natural densities and three levels of nutrient enrichment in triplicate 0.1 m³ mesocosms. Asterisks indicate significant ($p < 0.05$) effects of grazing.

Intermediate water-mixing rates are needed for optimal growth of submersed plants

Most coastal habitats are characterized by vigorous water circulation rates that influence diverse ecological processes, including the photosynthesis of aquatic grasses. Although there is ample evidence that water mixing is important in the functioning of aquatic plant communities (Madsen and Sondergaard 1983; Fonseca and Kenworthy 1987), most laboratory and mesocosm experiments have been conducted with little concern for appropriately scaling mixing rates. Typically, a rate of flow is arbitrarily chosen for a particular study, and no effort is made to consider possible artifacts resulting from unrealistic mixing regimes (Cornelisen and Thomas 2006).

MEERC scientists conducted experiments to investigate the effects of water circulation rates on growth of submersed plant communities established in mesocosms (tanks, 1.0 m³) (Zelenke 1999).*Vallisneria americana* was grown for 12 weeks in replicate mesocosms under three levels of mixing treatments as well as quiescent controls (Fig. 102). With increased turbulent mixing intensity, biomass of *V. americana* increased up to a threshold (Medium treatment). Further increase in turbulence (High treatment) caused a reduction in plant biomass (Fig. 103).

Apparently, diffusion-limitation across the boundary layer of *V. americana* leaves was overcome as mixing intensity was increased

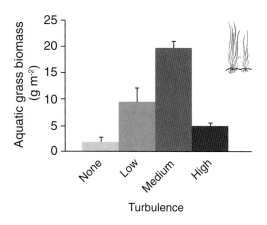

Figure 103: *Variations in biomass of* Vallisneria americana *in experimental ecosystems treated with a range of water mixing regimes from None to High (after Merrell 1996).*

from None to Medium treatments. Overall, the lowest biomass occurred in the quiescent (None) treatment, corroborating field observations that this species requires water circulation for healthy growth. Periphyton biomass also increased linearly from None to High mixing treatments. Damage to *V. americana* by physical stress and possibly wash-out of carbonic anhydrase were hypothesized to trigger the decline in plant production in the High treatment. This MEERC experiment indicates that flow and mixing regimes are important considerations in the design of mesocosms for submersed aquatic grass experimental studies.

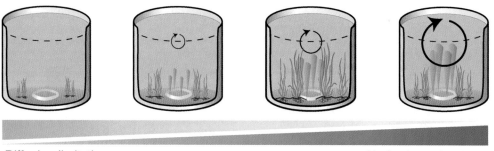

Figure 102: *Simplified representation of design for an experiment to study submersed plant (*Vallisneria americana*) community responses to water mixing intensities ranging from quiescent systems with diffusion limitation to vigorously mixed systems with potential physical stress. Note that the blue plumes rising from the silver rings depict bubble release used to mix experimental ecosystems.*

Biological factors: Marshes
J.C. Stevenson and K.L. Sundberg

Previous experimental marsh ecosystem studies have been limited in scope

Enclosed experimental ecosystems containing emergent marshes share some similarities with those containing submerged aquatic grasses and also differ markedly from pelagic-benthic mesocosms. While marsh mesocosms require many of the same considerations (e.g., light and sediment nutrients) as do those with submerged grasses, these vascular plants and underlying sediments are flooded periodically rather than continuously. They are also driven by both fresh and saline water fluxes with complex hydrology and hydrodynamics.

Although relatively few marsh mesocosm studies have been reported (Fig. 104), the growing enterprise of marsh creation projects for mitigation and wastewater treatment (Kadlec and Knight 1995) has generated useful knowledge about methods for successful construction of new marsh ecosystems. Most previous tidal marsh mesocosms were relatively small (<1 m², Fig. 104). A fundamental problem in marsh mesocosm design is the appropriate control of hydraulic inputs, which are critical in structuring marsh systems.

Marsh mesocosm experiments in MEERC began with 3 m long (1 m wide) mesocosms raised 30 cm at one end to create a gradient from high to low marsh zones. Groundwater was introduced by a diffuser at the high marsh, and

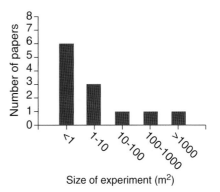

Figure 104: Most marsh mesocosm experiments have used small mesocosms.

tidal inputs and outputs were exchanged with an intertidal tank (30 × 60 × 10 cm) at the low end. Although these modifications were useful in studying groundwater effects on sediment redox, the relatively small size of these mesocosms precluded long-term biomass sampling.

A question addressed in MEERC marsh studies was the functional role of plant diversity under high and low nutrient inputs. A 2 × 2 factorial design was used with triplicate treatments. Based on previous experience and literature reports (Kitchens 1979), a 6 m² mesocosm was considered the most feasible and cost-effective size. For durability we used gel-coated fiberglass tanks (1 m × 6 m), placed at a slope of 1:20 (Fig. 105).

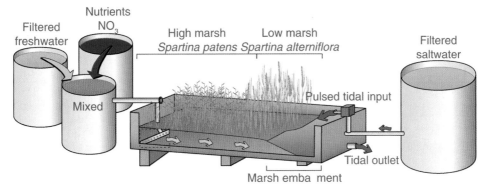

Figure 105: Diagram of MEERC marsh mesocosm experiments that explored plant species diversity under high and low nutrient inputs.

There are trade-offs between short-term and long-term experiments in experimental marsh ecosystems

In marsh mesocosm studies with turf cut from a marsh and reassembled in mesocosms, a piece of nature is isolated from normal inputs. Thus, it is advantageous to use mesocosms for short-term experiments and avoid long-term problems with artifacts. It is recommended that short-term marsh ecosystem experiments shoud use turf derived from natural marshes. MEERC researchers transported 20 × 20 × 20 cm marsh turfs by hand with little damage, but larger blocks had to be extracted and transported by mechanical means.

Marsh experiments in MEERC occurred outdoors, and freezing conditions in winter prevented tidal exchange throughout the year. Natural marsh systems are exposed to storm activity and ice events that scour and remove dead plant biomass from the previous growing season. In MEERC's long-term experiments, increasingly large amounts of dead biomass accumulated over the first few years, which may be viewed as an artifact.

In long-term experiments, various species become established independently in replicates, variance naturally increases, and statistical power declines. This is accentuated when small numbers of replicates are used for each treatment. Yet, long-term experiments may be necessary if the experimental questions involve changing species diversity.

Different methods for initiating marsh mesocosms may be appropriate depending on research questions (Table 10). When turf blocks are used, species diversity is not easily manipulated, and marsh communities are best created using seeds, sprigs, or sediment plugs to manipulate species composition. Although sterilized seeds are easily planted, seed collection and germination may be limited, suggesting that use of vegetative sprigs may be preferred.

Under conditions of high sediment fertility and organic content, experimental marshes may reach maturity in terms of ecological functions (e.g., nutrient retention, denitrification) within 3–5 years. When initial sediments have low organic matter and fertility, however, marshes may take a decade or more to optimize sediment functions. In long-term MEERC experiments that ran for >5 years, the soil organic content did not change, suggesting relatively rich and stable sediment conditions.

Table 10: *Advantages and disadvantages of the various methods for planting marsh grasses in mesocosms.*

		Advantages	Disadvantages
	Seeds	• High species and genetic diversity • Can be planted quickly • Deeper sediments needed • Can sterilize and control contaminants	• Long development time • Germination rates variable • Seed availability for certain species may be limited
	Sprigs	• Alcohol treatment to avoid contamination • Jump start on growth	• Bare-rooted material success is low • Labor-intensive planting
	Plugs	• High success rate • Natural microbial community	• Development time takes 2 years • Possible contamination (weeds, parasites)
	Swaths/turfs	• Fast development time • "Natural" plant-microbe community • Fewer artifacts • Strong root structure to start	• Less chance for new species • Hard to adjust plant density • Expensive, labor-intensive • Less responsive to treatment

Experimental marsh ecosystems need to be larger in multi-year studies

A fundamental question in marsh mesocosm design is, how large do experimental ecosystems need to be to address various research questions, particularly those that require long-term studies? When plant biomass is sampled destructively to obtain data on plant structure and tissue nutrient content, particularly as these change over annual cycles, this destruction may set a minimum size required for the study. Removal of too much plant biomass or sediments from the experimental system in too short a time interval can lead to serious experimental artifacts associated with altered plant growth and biogeochemical cycling.

In the MEERC marsh mesocosm studies, researchers used 0.1 m² quadrates (Fig. 106) for sampling aboveground biomass 3 times per year and with three replicates per zone. Thus, 1.8 m² of marsh plants per year were removed, and a total of 9 m² harvested over a 10-year study. If the total area harvested was limited to <10% of mesocosm, then an area of 90 m² would be required. Aboveground plant material harvested during the growing season tended to recover rapidly with newly formed plant material. MEERC researchers conducted simulation modeling experiments prior to establishing sampling protocols for future mesocosms studies. These simulations demonstrated that bi-monthly sampling of 5% of the plant biomass would result in negligible effects on marsh community productivity and biogeochemistry. Marsh plant harvest at higher percentages or frequencies led to disrupted functional ecology for the marsh system.

Destructive sampling of below ground biomass and sediments can create serious problems and associated experimental artifacts. MEERC studies used periodic core (0.02 m²) sampling (Fig. 106) to characterize sediment biogeochemical processes and below ground plant structure. Although holes can be filled with marsh sediment from other sources after core collection, it may take months or longer for that sediment to become colonized with natural flora and fauna and to establish balanced biogeochemical

Figure 106: Destructive sampling techniques used in MEERC marsh mesocosm experiments to obtain aboveground biomass included using 0.1 m² quadrates when removing above-ground biomass (top) and 0.02 m² cores for collecting sediment columns and belowground biomass (bottom).

gradients and processes. Routine coring in marsh mesocosm sediments 3 times per year in triplicate and in two zones would remove 0.36 m² sediment area per year, requiring a minimum of 4.5 m² marsh surface to maintain yearly sampling to <5% of the system area. Alternative non-destructive methods using porewater sippers and groundwater lysimeters are required for routine sampling for sediment biogeochemistry. When destructive coring is necessary, smaller diameter tubes should be considered.

MEERC marsh mesocosm studies used systems that were 2 m² for 2-year experiments, while 5 m² experimental marsh ecosystems were used in 5-year experiments. Most other mesocosms that have been used previously for marsh research have been much smaller (< 1 m², Fig. 104), with experimental durations of <1 year. Unfortunately, many geochemical characteristics evolve slowly over time as newly-created marsh systems develop. For example, marsh ecosystem restoration studies have shown that nutrient retention capacity may double over the first 5 years after marsh formation.

Light regime is an important factor in experimental marsh ecosystems

Light is a dominant factor in regulating the structure and function of all plant communities, including tidal marshes. Although some marsh mesocosms have been established indoor under artificial lighting, use of outdoors mesocosms maintained under natural sunlight avoids artifacts associated with reduced light quantity. Artificial lighting also generally fails to match natural diel patterns of sunlight. Even in greenhouses, wall and roof panels of varying chemical composition can cause spectral shifts that may alter natural patterns of stem elongation and flowering induction.

MEERC marsh mesocosms were located outdoors so that they received maximum natural radiation. Because the mesocosms were long and narrow to create a long tidal gradient, unexpected wall artifacts were encountered. The space between mesocosms allowed for ample growth of marsh grass at the mesocosms' edges. The grass often grew over the mesocosm walls, allowing plants at the edges to receive more light than those at the center of the mesocosms. Plants in the middle of the mesocosms were partially shaded by standing dead plant material, and live biomass was generally higher at the edges than in the center of the mesocosms (Fig. 107).

For natural temperate tidal marshes, much of the standing dead plant material is removed periodically by ice-rafting or scour from severe

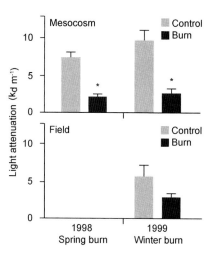

Figure 108: Light attenuation. Mesocosm high marsh (top panel), and field high marsh (bottom panel), 1998 and 1999. Error bars represent ± standard error of the mean. An asterisk indicates a significant treatment difference of 0.05 (Schmitz 2000).

storms. Fire is another natural (or human) disturbance that removes dead plant material, and MEERC researchers conducted burning experiments to investigate its effect on plant production and biomass. There was generally higher biomass at the edges compared to the center of experimental mesocosms (Fig. 107) in both burned and unburned portions of the mesocosms (Schmitz 2000). This edge-effect, which had not been previously reported, needs to be quantified more fully to aid in marsh mesocosm designs. Wider mesocosms (> 1 m) would reduce the problem, but might introduce other difficulties such as sampling the cross section.

In general, plant canopies in our marsh mesocosms were over-developed compared to those in nearby natural marshes. Shading under arguably an experimental artifact. Burning treatments significantly reduced this shading effect. However, light attenuation under unburned conditions was in the range of 9 m^{-1} in mesocosms compared to only 6 m^{-1} in the field (Fig. 108). Such dense shading at the experimental marsh surface inhibited growth of benthic algae, often a productive source of food and nutrient retention in many North American marshes (Gallagher and Daiber 1973).

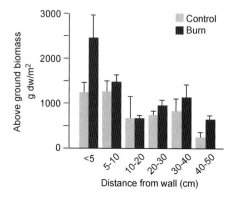

Figure 107: Mesocosm high marsh edge effects on aboveground biomass, in 1999. Error bars represent ± standard error of the mean. Burned values include the two mesocosms that received high nitrogen groundwater in 1999 (Schmitz 2000).

Marsh plant diversity can be easily manipulated in experimental ecosystems

The design and establishment of experimental marsh ecosystems must consider the question of what plant species need to be included to address particular scientific questions. Similarly, in the construction of marsh ecosystems for mitigation projects, designers need to have clear and reasonable targets for species composition and diversity needed to establish a functional ecosystem. Some of the earliest studies of stressed systems (Nixon 1969; Nixon and Oviatt 1973) indicated that as few as three plant species might be adequate for producing a functional salt marsh ecosystem. Tidal marshes are generally stressed plant communities that have far lower species diversity (often including monospecific zones) than adjacent uplands. Consequently, scientists need to ask how many marsh plant species need to be included in a mesocosm.

MEERC mesocosms were used to address this question. An experiment was conducted in 1995 using two levels of species richness established in 12 identical 6-m long fiberglass mesocosms (six replicates for each species level) filled with 30 cm of organic rich sub-tidal sediment from a nearby cove off the Choptank River. The Low Diversity (LD) treatment was initiated with plugs from natural marshes including three grass species. *Spartina alterniflora* was placed in the low marsh end (Fig. 109) of the mesocosm and *Spartina patens* and *Distichlis spicata* were placed in the high marsh end. The High Diversity (HD) treatment included the previously mentioned grasses plus two common sedges (*Schoenoplectus americanus* and *Eleocharis tenius*) and a malvaceaous species (*Kosteletzkya virginica*). The *E. tenius* and *K. virginica* did not survive the first winter, so *Eleocharis parvula* and *Hibiscus mosheutos* were added in the next spring. Although new recruits or volunteer plants were removed from the LD during the experiment, they were allowed to persist in the HD treatment.

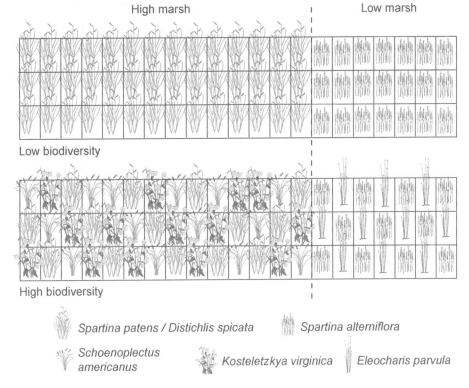

Figure 109: *Low biodiversity and high biodiversity treatments (six each) placed in 12 identical mesocosms were used to explore the number of plant species needed to establish functionality in mesocosm experiments.*

Marsh plant diversity is not critical for productivity or nutrient cycling

In the MEERC diversity experiment, species placement in marsh mesocosms was based on a randomized design in two blocks (i.e., high and low marsh; Fig. 110). The demarcation between high and low marsh was the high tide line in the mesocosm along the sediment surface. A range of variables were measured to assess marsh function. MEERC researchers used net primary productivity (NPP) of the marsh plant community and nitrogen buffering capacity (rates of N removal and denitrification). Three high diversity mesocosms received high nitrate inputs (by groundwater loading), and three low nitrogen control mesocosms received only filtered (reverse osmosis) groundwater and precipitation.

There were clear trends in productivity and nitrogen removal capacity that varied between mesocosms receiving High and Low nitrogen inputs. Although productivity and nutrient removal were slightly higher in HD mesocosms, differences were not significant when analyzed by a repeated measures ANOVA ($p = 0.05$). This supports the hypothesis that plant diversity does not appear to be critical to marsh ecosystems in terms of NPP and nitrogen buffering. However, questions remain concerning more subtle differences related to whether high diversity marshes might be better habitats for supporting greater abundance, and functional range for consumer species.

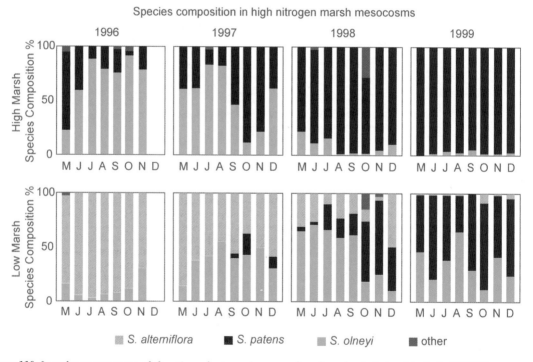

Figure 110: *In order to assess marsh function, plant species were placed randomly in two blocks in MEERC mesocosms. Species composition changed over the course of the experiment in both zones of these mesocosms receiving high nutrient loading.*

Successful marsh establishment depends on getting the hydrology right

It is well known that successful establishment of new marsh ecosystems as part of mitigation efforts depends on getting the hydrology correctly reproduced and scaled, particularly over the long term (Fig. 111). Although this is also true for the design and construction of enclosed experimental marsh ecosystems, many details remain to be resolved. For example, it is unclear how specific hydraulic regimes should be matched with particular marsh communities. Successful design of marsh mesocosms requires a basic understanding of plant requirements for water flow rates and seasonal distributions as they differ for various marsh species and soil conditions (Fig. 112).

In a sense, the highly regulated hydrological regime of the Florida Everglades is like an immense experimental ecosystem, where water diversions dating back almost a century in South Florida have caused portions of the Everglades to

Figure 111: Photograph of solenoid control on tidal water inflow in a MEERC marsh mesocosm.

dry out. Until recently, managers and researchers have had limited understanding of how much water marshes lose in evapotranspiration when functioning normally.

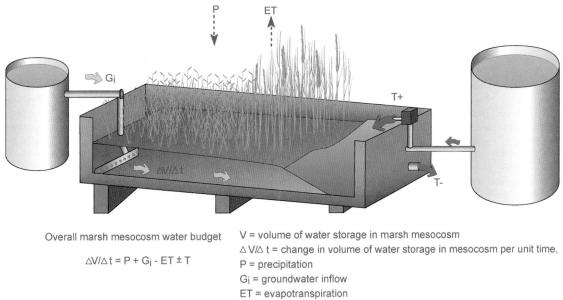

Overall marsh mesocosm water budget

$$\Delta V/\Delta t = P + G_i - ET \pm T$$

V = volume of water storage in marsh mesocosm
$\Delta V/\Delta t$ = change in volume of water storage in mesocosm per unit time,
P = precipitation
G_i = groundwater inflow
ET = evapotranspiration
T = tidal inflow (+) or outflow (-)

Figure 112: Understanding how evapotranspiration from the marsh surface influences the system's functionality has become increasingly important. By monitoring inputs and outputs of water in mesocosm experiments, researchers can estimate gaseous evapotranspiration.

Successful marsh establishment depends on getting the hydrology right

Using sensors, solenoids, and computer technology, large marshes as well as marsh mesocosms can be monitored and controlled closely with accurate water budgets at various scales (Fig. 113). Normally, researchers construct water budgets for flows of nutrients and sediment into and out of wetland systems. However, one of the important by-products of precise monitoring of inputs and outputs of water is an improved ability to estimate water loss through evapotranspiration from marsh plants and water or sediment surfaces. This is one of the strongest arguments for using mesocosms with an impermeable bottom when designing and conducting experiments.

For experimental tidal marsh ecosystems, pressure transducers that control tidal height need to be sensitive enough to measure millimeter scale changes in tide height and stable enough to tolerate seawater and outdoor conditions (Fig. 114). MEERC marsh mesocosms used a new low-head high-accuracy sensor built for tidal applications. Signals from the pressure sensors were logged into a data logger and a computer

determined the separate inflow to each mesocosm. It is important to note that solenoids can be damaged by periodic ice formation in winter.

Figure 113: MEERC shear turbulence resuspension Mesocosm (STORM) mixing and data collection center (p.205). This system controlled the mixing of non-repsuspension and resuspension mesocosms on the adjacent outdoor mesocosm pad.

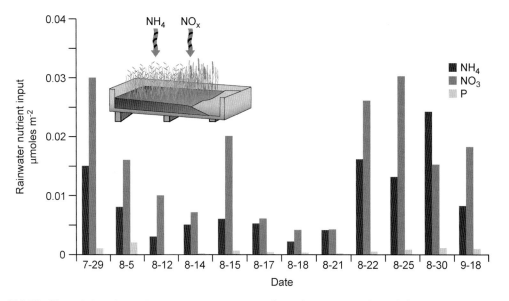

Figure 114: Weekly variations in nutrient sources to experimental marsh ecosystems through direct rainwater input.

References

Baker, J., R. Mason, J.C. Cornwell, J. Ashley, J. Halka and J. Hill. 1997. Final report to Maryland department of the environment, Ref. No. UMCES [CBL]:97–142.

Bartleson, R.D., W.M. Kemp and J.C. Stevenson. 2005. Use of a simulation model to examine effects of nutrient loading and grazing on Potamogeton perfoliatus L. communities in microcosms. Ecological Modelling 185:483–512.

Berg, G.M., P.M. Glibert and C.-C. Chen. 1999. Dimension effects of enclosures on ecological processes in pelagic systems. Limnology and Oceanography 44:1331–1340.

Boynton, W.R., J.H. Garber, R. Summers and W.M. Kemp. 1995. Inputs, transformations, and transport of nitrogen and phosphorus in Chesapeake Bay and selected tributaries. Estuaries 18(1B):285–314.

Brownlee, P.C. and F. Jacobs. 1987. Mesozooplankton and microzooplankton in the Chesapeake Bay. in S.K. Majumdar, L.W. Hall, Jr. and H.M. Austin (eds.). Containment problems and management of living Chesapeake Bay resources: 217-269. Pennsylvania Academy of Sciences, Easton.

Carpenter, S.R. 1996. Microcosm experiments have limited relevance for community and ecosystem ecology. Ecology 77:667–680.

Carpenter, S.R., J.F. Kitchell and J.R. Hodgson. 1985. Cascading trophic interactions and lake productivity. BioScience 35:634–639.

Chen, C.-C. and W.M. Kemp. 2004. Periphyton communities in experimental marine ecosystems: Scaling the effects of removal from container walls. Marine Ecology-Progress Series 271:27–41.

Chen, C.-C., J.E. Petersen and W.M. Kemp. 1997. Spatial and temporal scaling of periphyton growth on walls of estuarine mesocosms. Marine Ecology-Progress Series 155:1–15.

Chen, C.-C., J.E. Petersen and W.M. Kemp. 2000. Nutrient uptake in experimental estuarine ecosystems: Scaling and partitioning rates. Marine Ecology-Progress Series 200:103–116.

Cohen, J.E. and D. Tilman. 1996. Biosphere 2 and biodiversity: The lessons so far. Science 274:1150–1151.

Confer, J.L. 1972. Interrelations among plankton, attached algae and phosphorus cycle in artificial open systems. Ecological Monographs 42:1–23.

Cooke, G.D. 1967. The pattern of autotrophic succession in laboratory microcosms. BioScience 17:717–721.

Cornelisen, C.D. and F.I.M. Thomas. 2006. Water flow enhances ammonium and nitrate uptake in a seagrass community. Marine Ecology-Progress Series 312:1–13.

Cornwell, J.C. Unpublished data. MEERC report, U.S.E.P.A. Star Program.

Cowan, J.L.W. and W.R. Boynton. 1996. Sediment-water oxygen and nutrient exchanges along the longitudinal axis of Chesapeake Bay: Seasonal patterns, controlling factors and ecological significance. Estuaries 19:562–580.

Crawford, S.M. and L.P. Sanford. 2001. Boundary shear velocities and fluxes in the MEERC experimental ecosystems. Marine Ecology-Progress Series 210:1–12.

de Lafontaine, Y. and W.C. Leggett. 1987. Effect of container size on estimates of mortality and predation rates in experiments with macrozooplankton and larval fish. Canadian Journal of Fisheries and Aquatic Sciences 44:1534–1543.

Dudzik, M., J. Harte, A. Jassby, E. Lapan, D. Levy and J. Rees. 1979. Some considerations in the design of aquatic microcosms for plankton studies. International Journal of Environmental Studies 13:125–130.

Eppley, R.W., P. Koeller and G.T. Wallace, Jr. 1978. Stirring influences the phytoplankton species composition within enclosed columns of coastal sea water. Journal of Experimental Marine Biology and Ecology 32:219–239.

Estrada, M., M. Alcaraz and C. Marrasé. 1987. Effects of turbulence on the composition of phytoplankton assemblages in marine microcosms. Marine Ecology-Progress Series 38:267–281.

Fee, E.J. and R.E. Hecky. 1992. Introduction to the northwest Ontario lake size series (NOLSS). Canadian Journal of Fisheries and Aquatic Sciences 49:2434–2444.

Fonseca, M.S. and W.J. Kenworthy. 1987. Effects of current on photosynthesis and distribution of seagrasses. Aquatic Botany 27:59–78.

Fukuda, M.K. and W. Lick. 1980. The Entrainment of Cohesive Sediments in Freshwater. Journal of Geophysical Research 85(C5):2813–2824.

Gallagher, J.L. and F.G. Daiber. 1973. Diel rhythms in edaphic community metabolism in a Delaware salt marsh. Ecology 54:1160–1163.

Grice, G.D. and M.R. Reeve (eds.). 1982. Marine Mesocosms: Biological and Chemical Research in Experimental Ecosystems. Springer-Verlag, NY.

Gust, G. and V. Mueller. 1997. Interfacial hydrodynamics and entrainment functions of currently used erosion devices. Pages 149–174 in N. Burt, W.R. Parker and J. Watts (eds.). Cohesive Sediments. John Wiley and Sons, New York.

Harding, L.W., Jr., M.E. Mallonee and E.S. Perry. 2002. Toward a predictive understanding of primary productivity in a temperate, partially stratified estuary. Estuarine and Coastal Shelf Science 55:437–463.

Harte, J., D. Levy, J. Rees and E. Saegebarth. 1980. Making microcosms an effective assessment tool. Pages 105–137 in J.P. Giesy, Jr. (ed.). Microcosms in Ecological Research. National Technical Information Service, Springfield, VA.

Heath, M.R. and E.D. Houde. 2001. Evaluating and modeling foraging performance of planktivorous and piscivorous fish: Effects of containment and issues of scale. in R.H. Gardner, W.M. Kemp, V.S. Kennedy and J.E. Petersen (eds.). Scaling Relations in Experimental Ecology. Columbia University Press, New YorkPages 191–250.

Houde, E.D. Unpublished data. MEERC report, U.S.E.P.A. Star report.

Huettel, M. and A. Rusch. 2000. Transport and degradation of phytoplankton in permeable sediment. Limnology and Oceanography 45:534–549.

Kadlec, R.H. and R.L. Knight. 1995. Treatment Wetlands. Lewis Publishers, Boca Raton, FL.

Kemp, W.M., R. Batiuk, R. Bartleson, P. Bergstrom, V. Carter, G. Gallegos, W. Hunley, L. Karrh, E. Koch, J. Landwehr, K. Moore, L. Murray, M. Naylor, N. Rybicki, J.C. Stevenson and D. Wilcox. 2004. Habitat requirements for submerged aquatic vegetation in Chesapeake Bay: Water quality, light regime, and physical-chemical factors. Estuaries 27:363–377.

Kemp, W.M. and W.R. Boynton. 1984. Spatial and temporal coupling of nutrient inputs to estuarine primary production:

The role of particulate transport and decomposition. Bulletin of Marine Science 35:522–535.

Kemp, W.M., W.R. Boynton, J.E. Adolf, D.F. Boesch, W.C. Boicourt, G. Brush, J.C. Cornwell, T.R. Fisher, P.M. Glibert, J.D. Hagy, L.W. Harding, E.D. Houde, D.G. Kimmel, W.D. Miller, R.I.E. Newell, M.R. Roman, E.M. Smith and J.C. Stevenson. 2005. Eutrophication of Chesapeake Bay: Historical trends and ecological interactions. Marine Ecology-Progress Series 303:1–29.

Kemp, W.M., M.R. Lewis, J.J. Cunningham, J.C. Stevenson and W.R. Boynton. 1980. Microcosms, macrophytes, and hierarchies: Environmental research in the Chesapeake Bay. in J.P. Giesy, Jr. (ed.). Microcosms in Ecological Research p. 911–936. National Technical Information Service, Springfield, VA.

Kemp, W.M., J.E. Petersen and R.H. Gardner. 2001. Scale-dependence and the problem of extrapolation: Implications for experimental and natural coastal ecosystems. in R.H. Gardner, W.M. Kemp, V.S. Kennedy and J.E. Petersen (eds.). Scaling Relations in Experimental Ecology p. 3–57. Columbia University Press, NY.

Kemp, W.M., R.R. Twilley, J.C. Stevenson, W.R. Boynton and J.C. Means. 1983. The decline of submerged vascular plants in upper Chesapeake Bay: Summary of results concerning possible causes. Marine Technology Society Journal 17:78–89.

Kerhin, R.T., P.J. Blakeslee, N. Zoltan and R. Cuthbertson. 1988. The surficial sediments of the Chesapeake Bay, Maryland: Physical characteristics and sediment budget. Report of Investigation 48, Maryland Geological Survey.

Kitchens, W.M. 1979. Development of a salt marsh microecosystem. International Journal of Environmental Studies 13:109–118.

Kuiper, J. 1981. Fate and effects of mercury in marine plankton communities in experimental enclosures. Ecotoxicology and Environmental Safety 5:106–134.

Lewis, M.R. and T. Platt. 1982. Scales of variability in estuarine ecosystems. in V.S. Kennedy (ed.). Estuarine Comparisons. p. 3–20. Academic Press, NY.

Luckett, C., W.H. Adey, J. Morrissey and D.M. Spoon. 1996. Coral reef mesocosms and microcosms – successes, problems and the future of laboratory models. Ecological Engineering 6:57–72.

Luckinbill, L.S. 1973. Coexistence in laboratory populations of Paramecium aurelia and its predator Didinium nasutum. Ecology 54:1320–1327.

Madsen, T.V. and M. Søndergaard. 1983. The effects of current velocity on the photosynthesis of Callitriche stagnalis scop. Aquatic Botany 15:187–193.

Marvin-DiPasquale, M.C. and D.G. Capone. 1998. Benthic sulfate reduction along the Chesapeake Bay central channel. I. Spatial trends and controls. Marine Ecology-Progress Series 168:213–228.

Merrell, K.S. 1996. The effects of flow and mixing on Vallisneria and its associated community in experimental mesocosms. MS Thesis. University of Maryland, College Park, MD.

Mowitt, W.P. 1999. Scale-dependence of bay anchovy (Anchoa mitchilli) growth and top-down control by anchovies of plankton communities in estuarine mesocosms. MS Thesis. University of Maryland, College Park, MD.

Mowitt, W.P., E.D. Houde, D.C. Hinkle and A. Sanford. 2006. Growth of planktivorous bay anchovy Anchoa mitchilli, top-down control, and scale-dependence in estuarine mesocosms. Marine Ecology-Progress Series 308:255–269.

Muffley, B.W. 2002. Scale-dependent predatory effects of the Atlantic silversides, Menidia menidia, and lobate ctenophore, Mnemiopsis leidyi, on plankton communities in estuarine mesocosms. MS Thesis. University of Maryland, College Park, MD.

Murray, L., R.B. Sturgis, R.D. Bartleson, W. Severn and W.M. Kemp. 2000. Scaling submersed plant community responses to experimental nutrient enrichment. in S.A. Bortone (ed.). Seagrasses: Monitoring, Ecology, Physiology, and Management. p. 241–257. CRC Press, NY.

Naeem, S. and S. Li. 1997. Biodiversity enhances ecosystem reliability. Nature 390:507–509.

Neckles, H.A., R.L. Wetzel and R.J. Orth. 1993. Relative effects of nutrient enrichment and grazing on epiphyte-macrophyte (Zostera marina L.) dynamics. Oecologia 93:285–295.

Neundorfer, J.V. and W.M. Kemp. 1993. Nitrogen versus phosphorus enrichment of brackish waters: Responses of the submersed plant Potamogeton perfoliatus and its associated algal community. Marine Ecology-Progress Series 94:71–82.

Nixon, S.W. 1969. A synthetic microcosm. Limnology and Oceanography 14:142–145.

Nixon, S.W. and C.A. Oviatt. 1973. Ecology of a New England salt marsh. Ecological Monographs 43:463–498.

Odum, E.P. 1961. The role of tidal marshes in estuarine production. Conservationist 15:12–15.

Øiestad, V. 1990. Specific application of meso- and macrocosms for solving problems in fisheries research. in C.M. Lalli (ed.). Enclosed Experimental Marine Ecosystems: A Review and Recommendations p. 411–418. Springer. NY. Springer-Verlag, New YorkPages 136–154.

Parsons, T.R. 1982. The future of controlled ecosystem enclosure experiments. in G.D. Grice and M.R. Reeve (eds.). Marine Mesocosms: Biological and Chemical Research in Experimental Ecosystems p. 411–418. Springer-Verlag, NY.

Petersen, J.E., C.-C. Chen and W.M. Kemp. 1997. Scaling aquatic primary productivity: Experiments under nutrient- and light-limited conditions. Ecology 78:2326–2338.

Petersen, J.E., J.C. Cornwell and W.M. Kemp. 1999. Implicit scaling in the design of experimental aquatic ecosystems. Oikos 85:3–18.

Petersen, J.E., W.M. Kemp, R. Bartleson, W.R. Boynton, C.-C. Chen, J.C. Cornwell, R.H. Gardner, D.C. Hinkle, E.D. Houde, T.C. Malone, W.P. Mowitt, L. Murray, L.P. Sanford, J.C. Stevenson, K.L. Sundberg and S.E. Suttles. 2003. Multiscale experiments in coastal ecology: Improving realism and advancing theory. BioScience 53:1181–1197.

Petersen, J.E., L.P. Sanford and W.M. Kemp. 1998. Coastal plankton responses to turbulent mixing in experimental ecosystems. Marine Ecology-Progress Series 171:23–41.

Porter, E.T. 1999. Physical and biological scaling of benthic-pelagic coupling in experimental ecosystem studies. PhD Dissertation. University of Maryland, College Park, MD.

Porter, E.T., J.C. Cornwell, L.P. Sanford and R.I.E. Newell. 2004a. Effect of oysters Crassostrea virginica and bottom shear velocity on benthic-pelagic coupling and estuarine water quality. Marine Ecology-Progress Series 271:61–75.

Porter, E.T., M.S. Owens and J.C. Cornwell. 2006. Effect of manipulation on the biogeochemistry of experimental sediment systems. Journal of Coastal Research 22:1539–1551.

Porter, E.T., L.P. Sanford, G. Gust and F.S. Porter. 2004b. Combined water-column mixing and benthic boundary-layer flow in mesocosms: Key for realistic benthic-pelagic coupling studies. Marine Ecology-Progress Series 271:43–60.

Porter, E.T., L.P. Sanford and S.E. Suttles. 2000. Gypsum dissolution is not a universal integrator of "water motion". Limnology and Oceanography 45:145–158.

Purcell, J.E., U. Båmstedt and A. Båmstedt. 1999. Prey, feeding rates, and asexual reproduction rates of the introduced oligohaline hydrozoan Moerisia lyonsi. Marine Biology 134:317–325.

Reeve, M.R., G.D. Grice and R.P. Harris. 1982. The CEPEX approach and its implications for future studies in plankton ecology. in G.D. Grice and M.R. Reeve (eds.). Marine Mesocosms: Biological and Chemical Research in Experimental Ecosystems. Springer-Verlag, New YorkPages 389–398.

Resetarits, W.J., Jr. and J.E. Fauth. 1998. From cattle tanks to Carolina bays: The utility of model systems for understanding natural communities. in W.J. Resetarits, Jr. and J. Bernardo (eds.). Experimental Ecology: Issues and Perspectives. Oxford University Press, New YorkPages 133–151.

Ringelberg, J. and K. Kersting. 1978. Properties of an aquatic microecosystem: I. General introduction to the prototypes. Archiv für Hydrobiologie 83:47–68.

Roman, M.R. Unpublished data. MEERC report, U.S.E.P.A. Star report.

Roman, M., X. Zhang, C. McGilliard and W. Boicourt. 2005. Seasonal and annual variability in the spatial patterns of plankton biomass in Chesapeake Bay. Limnology and Oceanography 50:480–492.

Roush, W. 1995. When rigor meets reality. Science 269: 313–315.

Sanford, L.P. 1997. Turbulent mixing in experimental ecosystem studies. Marine Ecology-Progress Series 161:265–293.

Sanford, L.P. and S.M. Crawford. 2000. Mass transfer versus kinetic control of uptake across solid-water boundaries. Limnology and Oceanography 45:1180–1186.

Santschi, P.H., U. Nyffeler, R. Anderson and S. Schiff. 1984. The enclosure as a tool for the assessment of transport and effects of pollutants in lakes. in H.H. White (ed.). Concepts in Marine Pollution Measurements. p. 549–562. Maryland Sea Grant College, College Park, MD.

Schindler, D.W. 1998. Replication versus realism: The need for ecosystem-scale experiments. Ecosystems 1:323–334.

Schindler, D.E. and M.D. Scheuerell. 2002. Habitat coupling in lake ecosystems. Oikos 98:177–189.

Schmitz, J.P. 2000. Meso-scale community organization and response to burning in mesocosms and a field salt marsh. MS Thesis. University of Maryland, College Park, MD.

Short, F.T., D.M. Burdick and J.E. Kaldy III. 1995. Mesocosm experiments quantify the effects of eutrophication on eelgrass, Zostera marina. Limnology and Oceanography 40:740–749.

Steele, J.H. and J.C. Gamble. 1982. Predator control in enclosures. in G.D. Grice and M.R. Reeves (eds.). Marine Mesocosms: Biological and Chemical Research in Experimental Ecosystems. p. 227–237. Springer-Verlag, NY.

Sturgis, R.B. and L. Murray. 1997. Scaling of nutrient inputs to submersed plant communities: Temporal and spatial variations. Marine Ecology-Progress Series 152:89–102.

Tatterson, G.B. 1991. Fluid Mixing and Gas Dispersion in Agitated Tanks. McGraw-Hill, NY.

Twilley, R.R., W.M. Kemp, K.W. Staver, J.C. Stevenson and W.R. Boynton. 1985. Nutrient enrichment of estuarine submersed vascular plant communities. 1. Algal growth and effects on production of plants and associated communities. Marine Ecology-Progress Series 23:179–191.

Wiens, J.A. 2001. Understanding the problem of scale in experimental ecology. in R.H. Gardner, W.M. Kemp, V.S. Kennedy and J.E. Petersen (eds.). Scaling Relations in Experimental Ecology. p. 61–88. Columbia University Press, NY.

Zelenke, J. 1999. Tidal freshwater marshes as nutrient sinks: Nutrient burial and denitrification. PhD Dissertation. University of Maryland, College Park, MD.

Zieman, J.C., S.A. Macko, and A.L. Mills. 1984. Role of seagrasses and mangroves in estuarine food webs: Temporal and spatial changes in stable isotope composition and amino acid content during decomposition. Bulletin of Marine Science 35:380–392.

Tools for Design and Analysis of Experiments

The experiments discussed in the previous two section of this book have been explicitly designed to examine how ecosystem dynamics vary with scale. The results of these experiments clearly lead to the conclusion that scaling choices can profoundly affect experimental outcome.

Most mesocosm experiments are conducted in a single size and shape system and are designed to address questions other than the effects of scale—for instance, the effects of nutrients, toxins, or species manipulations on ecological dynamics. This section addresses rules, tools, and considerations relevant to design and interpretation that are available to all researchers to ensure that experiments conducted in mesocosms adequately account for the effects of scale.

Specifically, this section of the book considers (1) statistical considerations and the implications of variability in nature for experiments, (2) dimensional analysis as a tool for designing and interpreting experimental ecosystems, and (3) simulation modeling as an additional tool for design and interpretation.

Statistical considerations
R.H. Gardner

Statistical power increases with increasing number of replicates, which increases experimental costs

A variety of statistical tests can be used to assess ecological responses in mesocosm experiments, including t-tests, F-tests, analysis of variance (ANOVA), repeated measures analysis, and spectral analysis. In all cases, statistical power, the ability to detect differences among treatments, increases with increasing number of replicates (Fig. 1).

The Student t-test is one of the simplest statistical tests to perform, and provides insight into the nature of statistical comparisons. The t-test (in its simplest for m) is used to compare an observed or experimentally determined mean value from some perfectly known expected value (typically the *null hypothesis* of no treatment effect). In the following example, the expected value is 0 and provides a statistical evaluation of the null hypothesis that "no change due to experimental conditions exists." Hence:

$$t = [\, x - 0\,] \, / \, s/(n)^{1/2}$$

where s is the standard deviation of the observations made in replicate mesocosms and n is the number of replicate mesocosms.

All else being the same, the critical value of t verifying a statistically significant difference (e.g., less than a 5% chance, α, that this difference is due to random error) declines rapidly as sample size (i.e., number of replicate mesocosms) increases. Figure 1 illustrates a case where a mean difference of 0.5 has been observed. If the variance is equal to 1.0 then a sample size, n, must be greater than 12 for this level of difference to be statistically significant. Further increases in the number of replicates will increase our statistical power to detect significant difference, but will also increase the expense associated with this experiment.

The *null hypothesis* of no effect is difficult to reject because it is estimated without error (i.e., it is constant). Statistical power to reject the null hypothesis increases with decreasing variability (Fig. 2) as well as with increasing replication.

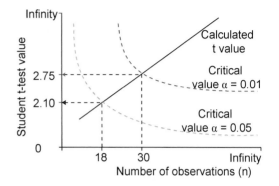

Figure 1: *The ability of statistical tests to resolve smaller and smaller differences increases as sample size (e.g., number of replicate mesocosms), n, increases. This example assumes that the difference between the mean value observed and the expected value is equal to 0.5 and the standard deviation is equal to 1.0. If these two statistical parameters are fixed over all sample sizes, then the value of the Student t will increase in value as a function of the n^{1/2}, as illustrated. When n = 18, then the t equals 2.1 and our statistical confidence, α, equals 0.05. At this level of confidence there is less than 1 chance in 20 that the observed difference is due to the random effects. Further increases in sample size elevate the value of t and statistical confidence levels also increase. At n = 30 the t value is 2.75 with an α of 0.01 (1 chance in 100 of random effects producing observed results).*

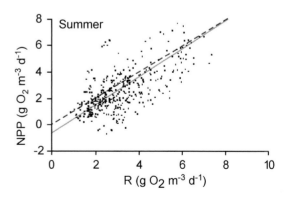

Figure 2: *In these data, regressions of net daytime productivity (NPP) versus night-time respiration (R, solid line) were not significantly different from the expected slope value of 1.0 (dashed line) with increased variability, (scatter) results would no longer be statistically significant (Petersen et al. 1997).*

Analysis of variance ANOVA is used when comparisons are made among multiple treatment groups

The t-test is a useful and powerful tool for making comparisons between a single data set and a constant (i.e., 0) or between two data sets (Fig. 3). In either case the question addressed concerns the likelihood that an observed difference is due to chance alone. Because there can be no absolute answer to this question, the response is phrased in terms of a probability, α. When α is small (usually ≤ 0.05) one is confident that observed differences are unlikely to be due to chance alone and that repeating the experiment will give similar results.

But what happens when one wants to make more than one comparison? For instance, the MEERC pelagic mesocosms came in five configurations (p. 36). The objective was to perform experiments measuring differences in ecosystem dynamics resulting from these different configurations. It is possible to use multiple t-tests to compare the five mesocosms types, but this is not advisable. The reason is that ten pair-wise comparisons will be needed to fully compare all five mesocosms. As the number of comparisons, n, increases, one's confidence $(1-\alpha)$ that differences that are identified as significant are not due to random effects declines. If $\alpha = 0.05$ for a single comparison, then with ten comparisons, $\alpha = (1-0.95^{10}) = 0.41$. This means that in experiments with five treatment groups, pair-wise comparisons

will find a significant difference by chance alone in a large percentage of the cases. This problem is so familiar that statisticians have labeled it as a Type I error.

Analysis of Variance (ANOVA) avoids the problem by the simple device of pooling all observations and making a single, global test of differences among treatments. This test, called the F-test, is the ratio of the mean squared difference (variance) between treatments compared to the mean squared difference among all observations. Because the distribution of the F-test is known, a probability that the results are non-random can, like the t-test, be estimated. When this probability is less than α, then one can be confident that real differences are likely to exist.

ANOVA is well-suited to complex experimental designs that simultaneously manipulate multiple factors (e.g., light, temperature, nutrients, etc.). If the experiment has been properly designed, ANOVA can identify the variance associated with each factor (and interactions among factors), and quantify the strength of their effects. A detailed discussion of this topic is beyond the scope of this book. Interested readers should consult comprehensive statistical texts for more information. However, one advanced technique— repeated measures design—is of particular interest to mesocosm studies.

Figure 3: The experimental systems shown in case a did not respond to the addition of nutrients. Consequently, the variability in the observed data completely overlapped the no effect case. Case b shows that only a small portion of the information overlaps 0 with the measurable effect (i.e., the difference between the dotted and solid line) exceeding the confidence of $\alpha < 0.05$.

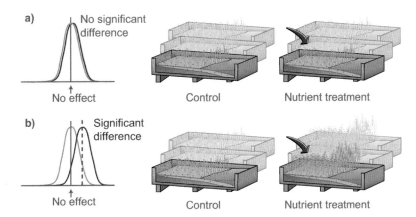

Repeated measures can be used to increase the statistical power of ecological experiments

A key problem in ecological studies is that an experimental unit—be it field plots, stream reaches, or mesocosms—often will have or will rapidly develop unique properties that distinguish it from other replicate units. For instance, the spatial location and disturbance history of field plots will affect species composition and ecosystem productivity. As an example, even though uniform procedures were used to initialize and maintain the MEERC pelagic-benthic mesocosms, no two systems were ever exactly alike. Because small differences are often amplified with time, the variance among replicate experimental units may also grow through time.

The unique attributes of similar experimental units can be considered within ANOVA (Fig. 4) by using a repeated measures design. With this approach, changes in sequential measurements of a single unit are used to characterize the variability of that unit. Essentially, instead of comparing the absolute value of some variable such as chlorophyll among replicates and among treatments, the researcher compares changes over time in chlorophyll (e.g., the increase in time of chlorophyll between days 1 and 3 in each mesocosm). Repeated measures address two questions: do individual treatments change significantly over time and do changes over time differ significantly among treatments? Measuring change often allows the researcher

Figure 4: *Multiple treatment effects, such as ecosystem size, shape, and nutrient enrichment, can be analyzed using ANOVA. This approach has been the principle statistical technique used in MEERC experiments to detect multiple treatment effects, and their possible interactions. Generally, three replicates have been included per treatment.*

to detect treatment effects even when replicates have very different mean values (see Crowder and Hand 1990 for details on repeat measures analysis).

Berg et al. (1999) used a repeated measures design to study the changes in nutrients and chlorophyll concentration in five pelagic-benthic mesocosms of different sizes and shapes (Fig. 4). The experiment involved three replicates of each of five mesocosm types measured at eight sequential times, giving a total of 120 observations. The repeated measures design controlled for the unique attributes of each replicate, and allowed researchers to conclude that both mesocosm dimensions and time had significant effects on nutrient concentration and chlorophyll *a* (Table 1).

Table 1: *Example of repeated measures ANOVA. Researchers assessed the effects of mesocosm type, time, and interactions between type and time on nutrient concentration (Berg et al. 1999).*

Source of variation	Significance levels (a)			
	Silica	NH_4^+	NO_3^-	Chl *a*
Mesocosm type	0.01	0.0005	0.0001	0.005
Time	0.0001	0.0001	0.0001	0.001
Mesocosm type x Time	0.0001	0.0001	0.0001	0.06

The world is variable
R.H. Gardner

Variability is an important consideration in experiments

It is crucial for the experimentalist to recognize and address the fact that components of the biotic world are never static, but are constantly adjusting to variable environmental conditions. Some of this variability is familiar and easily described. For example, weather patterns vary daily, light and temperature vary with seasons, and year-to-year changes in the frequency and intensity of storm events (Fig. 5) have been recorded for centuries. Although patterns of change are repeated over fixed intervals of time (e.g., seasons) or space (e.g., undulations in the land surface), other changes occur in aperiodic patterns. Our familiarity with aperiodic events led to a rich vocabulary describing these events as *random, stochastic, chaotic,* or simply *chance events.*

In nature, ecosystem productivity and persistence depend on the ability of biota to adapt to internal and external sources of variability as well as to trended environmental change. For example, flowers and leaves emerge in a predictable response to changes in light and temperature in spring, but are sometimes killed by unexpected frosts. Certain species have evolved to take advantage of disturbance events. Other species, that lack adaptations to survive this change, must seek more stable habitats. For example, if the frequency and severity of frost events increases, poorly adapted plants will experience local extinction.

Variability is ubiquitous, and understanding the source and the response of ecological systems to variability is a crucial challenge. There are three general approaches to this addressing variability:

(a) Frequent long-term measurements may be employed in nature to estimate the average (expected value) and the degree to which deviations from the average occur;

(b) Controlled experiments may be performed that reduce or eliminate random and periodic variability so as to maximize statistical power to detect treatment effects.

(c) Controlled experiments can be designed to assess the effects of variability by systematically manipulating the status of environmental factors that are known to vary in nature.

a) Current speed in nature

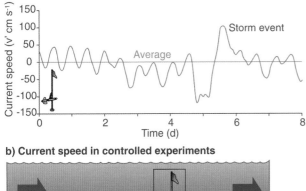

b) Current speed in controlled experiments

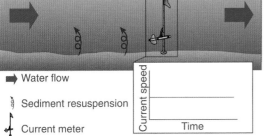

Figure 5: Periodic and aperiodic variability in physical conditions within natural aquatic environments are often driven by external factors associated with seasonality and storm events. (a) The strength and direction of tidal flow exhibits regular periodicity (with these hypothetical data) in response to lunar cycles and basin morphology on scales of days and months. Storm events can also exhibit seasonal periodicity, but are more variable. (b) Experiments conducted in mesocosms allow researchers to control the degree of variability. Thus, variability can either be held constant to evaluate effects of average conditions, or otherwise manipulated to evaluate the effects of different patterns of variability.

Understanding the relationship between variability and expected values is a key objective of statistics

Statistical theory tells us that, as the sample size increases, the average of repeated random events will eventually converge to an *expected value*. Figure 6 illustrates this principle with a numerical experiment. Samples of random numbers are drawn from a uniform distribution (expected value = 50, range between 0 and 100). The mean value is repeatedly estimated as sample size increases. Repeated sampling results in the convergence to the expected value. Only 10–12 samples are usually needed to adequately estimate the mean value if the variance of events are symmetrically distributed around the average value.

Statistical theory indicates several other important features of these results:

• Estimates of the mean are repeatable if all samples are independent;

• The mean value is an unbiased estimator of the expected value—that is, small samples are as likely to be over-estimates as they are to be under-estimates of the true average;

• The precision of the estimate of the average value increases as sample size increases.

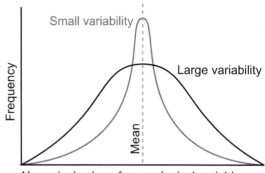

Figure 7: Whether populations have a small or large variability, the sample mean is a suitable estimate of the expected value. However, variability in an ecological variable can be as (or more) important as the mean in determining ecological dynamics. Variability must be carefully considered and controlled in experimental ecosystems.

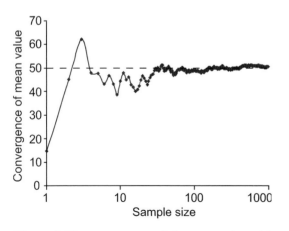

Figure 6: The convergence of the mean value with increasing sample sizes. The precision of estimates of the expected value (in this case) increases with increasing sample size.

Parameters of normal distribution

The mean, $\bar{x}$, is the average of all observations, x_i, while the variance, v, is the squared deviation of observations from the mean divided by the number of observations (n) minus 1 (Fig. 7).

The standard deviation, s, is one of the most widely used measures of variability and defines the degree of spread of observations away from the mean. The standard error, $s.e.$, estimates the precision with which the mean is estimated.

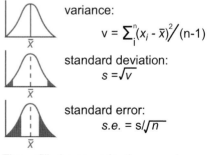

variance:
$$v = \sum_{i}^{n}(x_i - \bar{x})^2 / (n-1)$$

standard deviation:
$$s = \sqrt{v}$$

standard error:
$$s.e. = s/\sqrt{n}$$

The unfilled area under the curve in each graph above corresponds with the three measures of variability.

Variability should be controlled in experimental ecosystem for precise, repeatable measurements

Unbiased, precise, and repeatable measurements are the universal objective of scientific research. This goal often requires that variability be quantified and controlled to reveal subtle responses of biota to changing conditions. Experimental systems achieve this objective in three important ways:

First, the aperiodic, and often the periodic, variability of environmental variables (e.g., light, temperature, salinity, etc.) is reduced or eliminated by the careful construction of environmental controls (Fig. 8). Lights are designed to provide constant on-off illumination, heaters or coolers maintain temperature within tight bounds, water quality is monitored and adjusted to keep conditions within a specified range, etc.

Second, replicate systems are used for controls and treatments to provide a sample size that is large enough so that averages among replicates genuinely characterize response and responses to treatment can be statistically detected. Experiments may need to be repeated multiple times to achieve a level of statistical power needed to detect these effects.

Third, a sampling regime is designed to characterize variability and mean conditions within mesocosms. The frequency of sampling must adequately capture changes over time. If the environment within the experimental ecosystem is heterogeneous, samples must also be taken at multiple locations in space. For this reason, mesocosms are often designed to maintain an internally homogeneous environment so that samples taken in a few locations will adequately characterize overall environmental conditions.

Statistical power is maximized when experimental designs minimize the variability among replicates through tight regulation of environmental conditions. A fundamental problem with this is that nature is genuinely variable and minimizing variability by holding conditions constant potentially distorts ecological dynamics and the conclusions that researchers draw. The following pages consider how variability can be quantified and the implications of variability on the design and interpretation of experimental ecosystems. Ultimately a compromise must be struck so that the variability included in each experiment provides an adequate picture of real ecosystems while also controlling the overall level of variability considered.

Figure 8: The natural variability of environmental conditions is controlled and reduced in an experimental system.

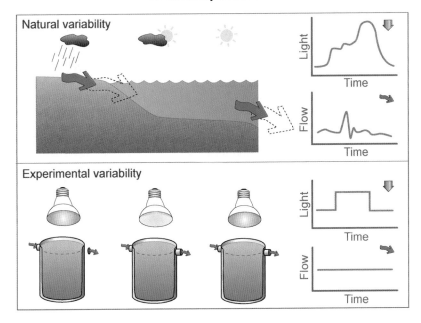

Periodicity often drives the variability of natural systems, can be quantified, and is a critical consideration in experimental design

Periodic phenomena abound in nature. When repeated measures are sequentially made within systems that are characterized by periodicity, observations will not be independent of one another, resulting in a biased estimate of statistical parameters (e.g., the mean and variance). Special methods such as autocorrelation and spectral analysis can be used to characterize patterns in

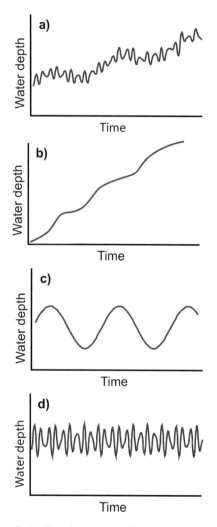

Figure 9: *(**a***) *Complete pattern observed is the sum of the three components of variability depicted in panels b, c, and d. (**b***) *Long-term trend (i.e., sea-level change). (**c***) *Periodic variation due to the lunar cycle. (**d***) *Short-term changes due to the tidal cycle. Time scales in panels decrease: b > c > d.*

periodic variability in physical and biological phenomena (Platt and Denman 1975).

Figure 9 is an exaggerated illustration of the decomposition of variance into three independent components. Water depth in a hypothetical tidal basin (Fig. 9a) is affected by long-term trends of increased water levels due to rise in sea-level (Fig. 9b), lunar effect on tidal height (Fig. 9c), and period oscillation of water levels due to tidal change (Fig. 9d). Factors in panels c and d are periodic in nature, but the factor in panel b is a non-periodic trend within the range of the measurements.

Spectral analysis identifies periodic changes in measurements by regressing the variance (power or spectral density) against possible frequencies (or spectra) of periodicity. Because non-periodic phenomena (Fig. 9b) may interfere with the spectral analysis, trends are removed before analysis proceeds. Then, frequencies most responsible for the observed change in variance are identified.

This discussion of variability is relevant to experimental ecology because one of the goals of experimentation is to control variability in order to maximize statistical power. A fundamental dilemma for experimentalists is that species have evolved and ecological systems have developed in response to periodicity and trends in the external factors that influence ecological dynamics. On one hand, removing and simplifying periodic and aperiodic variability is necessary for successful detection of cause and effect. On the other hand, this distortion of variability inevitably alters ecological response dynamics. Too little attention has been given to this problem and researchers generally assume the *mean-field approximation*, that is, that simulating conditions that are variable in nature with average values in experiments is a reasonable approximation. Techniques like spectral analysis can be used to better characterize variability in nature and then used to design experimental systems that simulate the critical aspects of variability so that these experiments more accurately represent nature.

Spectral analysis reveals scale-dependent changes in data that can inform experimental design

Spectral analysis is a statistical method used to describe how variability changes with scale. To illustrate a spectral analysis, the rate of water flow out of the major rivers of Chesapeake Bay (Susquehanna, Potomac, Patuxent, Choptank, and James rivers) was analyzed to produce average normalized spectral densities (Pandey et al. 1998). Plotting the logarithm of the variance (or the spectral density, Fig. 10) against the inverse of frequency provides an estimate of the change in variance with scale. Note that the time scale is inverted on the x axis in spectral density plots. That is, the higher frequencies (daily events) are plotted on the right side of the x axis, while lower frequencies (annuals) are plotted on the left side. The slope of this relationship, β, defines the ranges over which linear dependencies, and the ability to make simple extrapolations, exist. Three dominant patterns are evident in Fig. 10:

1. Annual variation in river flow is the dominant factor affecting the variability of river input to Chesapeake Bay. Year-to-year variation is more important than any other periodic factor.

2. Except for this annual peak, the spectral density increases linearly with decreasing frequency ($\beta = 0.67$ from 14 days to decadal time-scales).

Lower frequencies (also referred to as long-wave patterns or red noise) often explain most of the variability of physical phenomena, such as water flow.

3. A different pattern of variances is associated with rapidly changing events (i.e., those that occur with high frequencies, ~7–10 days). In this region, there is a steep linear trend ($\beta = 2.54$) attributable to periodic shifts in weather systems (Pandey et al. 1998). Although important, these higher frequencies explain far less of the overall variance than do long-wave patterns.

Mesocosm experiments may run for days or years (Petersen et al. 1999) and span the spectral shift in variability. Minimizing external sources of variability in experimental systems increases statistical power to test treatment effects, but may distort dynamics. Mesocosms can test the effects of different patterns of variability on ecological dynamics and potentially improve future experimental designs. For example, mesocosm experiments have tested the effects of treatment factors (e.g., nutrient loading) administered at different frequencies (Petersen et al. 1999). Such experiments help to define potential artifacts of controlled studies where external variability is minimized.

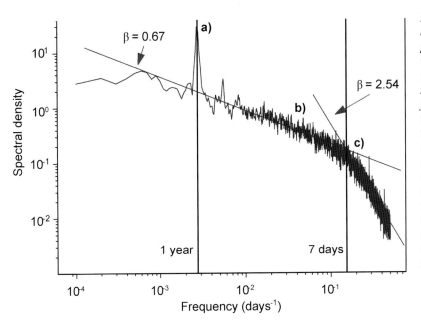

Figure 10: The spectral analysis of river flow data illustrates the predictable pattern of change in variance of riverine input to the Chesapeake Bay with time. (**a**) A peak in variance occurs at a frequency of 1-year, due to year-to-year differences in rainfall (i.e., wet and dry years). Changes at shorter time scales are predictable (constant slope of $\beta = 0.67$) until (**b**) a change in periodicity associated with weather fronts is encountered (7–10 day period). (**c**) Although weather fronts are important, they explain much less of the total variance than year-to-year differences in riverine runoff.

Prior research provides lessons about statistics and scale-dependent phenomena that informs experimental design

Theoretical and empirical research on variability has resulted in a variety of lessons that are relevant to experimental ecology:

• Spectral analysis (see box) is one member of a family of statistical methods for detecting scale dependencies. Other methods include autocorrelation, regression methods, and estimates of semivariance. These methods are all based on the analysis of a change in variance with scale.

• Failure to recognize long-term trends and periodicities will result in biased estimates of the mean (Fig. 11) and variance and prevent results from being meaningfully extrapolated across temporal or spatial scales.

• Short-term can only be used to characterize short-term phenomena ($\beta = 2.54$, Fig. 10), while multiple long-term studies must be conducted in order to characterize year-to-year changes in environmental variables.

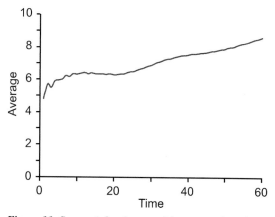

Figure 11: *Sequential estimates of the mean value of time series data that contain a trend. The presence of trends results in a mean value that increases with time.*

Data requirements and rules for spectral analysis

• A data set composed of long-term, continuous measurements;

• An uninterrupted sampling frequency (i.e., hourly, daily, etc.). The frequency of samples defines the resolution of the data while the total number of samples defines the extent of the data;

• Removal of long-term trends before analysis (Fig. 9b);

• The smallest scale of periodic phenomena that can be identified is equal to 2 times the data resolution, while the longest frequency is equal to half the length of the data record; and

• Spatial data may be analyzed in a similar manner to time data.

• Statistical methods, such as spectral analysis, can provide insights into the range of scales over which observational information may be used to forecast future conditions. Although non-linear relationships can be extrapolated, the presence of a linear relationship makes extrapolations particularly easy.

• The relative amount of variance explained by a particular process is a direct measure of its importance for extrapolation. Thus, year-to-year changes in river flow are much more important than weekly or daily fluctuations in river flows.

• The reliability of scale-dependent extrapolations are dependent on the adequacy of the model and the data upon which the extrapolations are based. Because models and data are developed from past experiences, new conditions or unexpected phenomena may alter ecosystem dynamics, thus requiring a reassessment of past predictions. The iterative process of model development, system measurement, and evaluation of predictions must continue until all possible outcomes have been described.

Spectral analysis can be used to predict spatial patterns and enhance extrapolation of experimental results

Spectral analysis is a statistical model of the change in variance with scale. The statistical equation is estimated from spatial or temporal data by regression methods. Therefore, the equation may be inverted to predict temporal and spatial patterns of change. An example of such a prediction is shown in Fig. 12. The two-dimensional image in the top panel represents the expected spatial pattern of chlorophyll normally observed in Chesapeake Bay in August. Analysis of actual chlorophyll patterns (Fig. 12, middle and bottom panel) for the Patuxent River had shown that spectral slope was nearly constant at $\beta = 2.0$ across a range of spatial scales from meters to kilometers. This equation estimated from spatial sampling of chlorophyll concentrations was then inverted and underwent the following manipulations:

- A normalized (mean = 0.0, standard deviation = 1.0) two-dimensional surface (30 km x 30 km) was produced.

- The normalized surface was multiplied by the observed variance associated with actual estimates of chlorophyll levels in August in Chesapeake Bay.

- The observed mean value was then added to all points to produce one possible realization of the spatial distribution of chlorophyll. Figure 12a compares pattern generated by spectral analysis with measured chlorophyll patterns in the Patuxent estuary, Figure 12b.

This general process is one example of a spatial extrapolation using the empirical model derived from spectral analysis to generate patterns of variability of environmental variables. Other variables that are often measured across spatial scales (e.g., temperature) can be extrapolated similarly, although the slope of these relationships, β, must be independently estimated. Because the variability of environmental drivers has been ESTIMATED independent of mesocosm experiments, spectral analysis allows one to define the range and variability of physical phenomena

that drive natural systems. Such information is a prerequisite for extending mesocosm models from the confines of enclosures to the spatial extent of natural systems. It is also possible to use this information to drive experimental systems, releasing experiments from the restrictions of constant light and temperature regimes that are normally employed. A rich area of investigation awaits the experimentalist who wishes to investigate biotic response to frequency-dependent changes in environmental conditions.

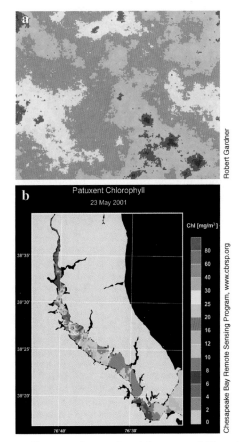

Figure 12: (*a*) *Prediction of spatial patterns of chlorophyll concentration based on a spectral inversion with* $\beta = 2.0$. (*b*) *Spatial patterns of chlorophyll concentrations measured in Patuxent River on 23rd May 2001 are similar to those predicted using spatial regression methods.*

Experiments and statistics are necessary but not sufficient conditions for predicting and interpreting nature

The main goal of manipulative experiments is to sufficiently understand the dynamics of natural systems so that their response to changing environmental conditions can be predicted. The statistical analysis of experimental studies is an important step leading to this understanding. However, experiments and statistics alone are insufficient for prediction, as illustrated by the following three points:

• Statistical significance and ecological importance are not equivalent. The formula for the t-test shows that the power to resolve small differences increases as the square root of the number of replicates increases ($n^{1/2}$). This means that sufficiently high replication can be used to resolve very small differences. It is possible that the magnitude and ecological importance of the effect may be very small even though the statistical significance is high.

• Experimental ecosystems are necessarily simplifications and abstractions of nature. Statistical power comes, in part, from simplifying the organisms, habitat, physical forcing, and variability present in the experiments relative to what is present in the natural world. Because experimental systems therefore are different from the natural ecosystems that they represent, researchers cannot be completely confident that treatment effects observed in mesocosms are actually operating in nature. A variety of artifacts associated with enclosure can also distort dynamics and response to treatments.

• Experimental ecosystems are a truncated sample of spatial and temporal scales. Experiments must concentrate on a limited set of factors. Other variables, important to natural systems, are eliminated in order to perform a controlled study. Using the concepts of spectral analysis, this process is equivalent to a truncation of frequencies. For instance, the effects of tides, storms, and natural disturbances are usually eliminated from experimental studies. Even when natural systems are studied (i.e., whole-lake manipulations), examining responses on the order of decades and centuries are impractical.

Consideration of the missing frequencies (Fig. 13) must occur in order for the results of experimental studies to be extrapolated to the spatial and temporal scales of natural systems.

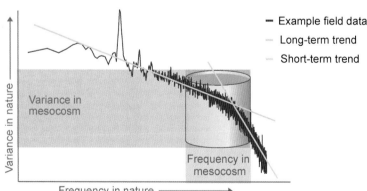

Figure 13: The range of frequencies measured in experiments is often a limited subset of the temporal frequencies affecting natural systems. Linear extrapolation based on mesocosm results will often fail to predict the trends of change affecting actual systems. Thus, extrapolation from the scales of mesocosms to natural systems may have considerable error unless the effects of new, scale-dependent processes are considered. Spectral analysis helps to identify the range of scales over which extrapolations are reliable (Fig. 10).

Dimensional analysis
J.E. Petersen and W.M. Kemp

Dimensional analysis is a tool for scaling experimental ecosystems

A guiding principle in designing scale-sensitive studies is to maximize consistency among three sets of scales: (1) scales associated with the research goals, (2) scales associated with the organisms and processes under investigation, and (3) scales associated with the physical conditions of the experiment (Petersen and Hastings 2001; Petersen and Englund 2005).This means intentionally designing the experiment to create a match between experimental and characteristic scales (see p. 18 for definitions of observational, experimental, and characteristic scale). For instance, for many studies it is appropriate to choose the duration of the study (the experimental extent) to be a whole integer multiple of the generation times of the dominant organisms (the characteristic extent). This practice of intentionally matching the scales of experimental design with the scales of the organisms and processes under investigation ensures that change observed during the experiment is not biased by starting and ending at different phases in an organism's life cycle. In addition to generation time, it is important to consider scaling characteristics such as the home range (i.e., a characteristic spatial extent) of the dominant organism in selecting mesocosm size (i.e., the experimental extent). Experimental designs that fail to match the experimental scales with the characteristic scales of key organisms and processes frequently result in erroneous conclusions (Tilman 1989).

A scaling distortion occurs when there is a mismatch between observational, characteristic, or experimental scales. Because mesocosms are smaller than natural ecosystems, are operated for brief periods relative to natural processes, and generally contain reduced biological diversity, time and space scales are distorted relative to nature. A key question is, can researchers compensate or adjust for these scaling distortions in the design and interpretation of experiments so that key ecological relationships are maintained and meaningful conclusions can be drawn about the natural ecosystems represented? This can be accomplished by dimensional analysis, the use of a variety of techniques based on the proposition that certain universal relationships should apply regardless of the dimensions of the system under investigation (Petersen and Englund 2005; Tilman 1989). These techniques have been widely used in engineering to design scale models that retain critical properties of larger systems. The premise is that miniaturizations and scaling distortions in one variable might be counter-balanced by intentional adjustments to other variables so as to conserve desirable relationships (Fig. 14).

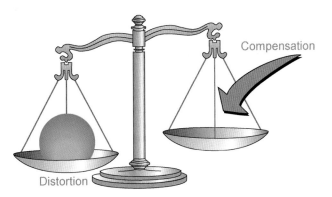

Figure 14: *Relationships among organisms and processes are inevitably distorted in the construction of small-scale experimental ecosystems. Compensatory adjustments in experimental design and interpretation can be used to account for these distortions and to extrapolate results to nature.*

Non-dimensional variables are scale independent

A dimensionless expression or non-dimensional variable is simply a mathematical expression in which terms are combined in such a way that the units cancel. For example, commonly used non-dimensional variables include percentages (part/whole), the coefficient of variation (standard deviation/average), and many other simple ratios (e.g., reaction rate/dispersal rate).

An important property of non-dimensional variables is that they are scale independent; their magnitude is independent of the particular units used or physical dimensions of the system. To understand how and why this is important, consider photosynthesis, respiration, and secondary consumption in two environments, marsh and open water. In the marsh, it might be reasonable to express all three variables in

units of $g\,C\,ha^{-1}\,month^{-1}$, whereas one might use $mg\,O_2\,L^{-1}\,day^{-1}$ for these variables in the planktonic system (Fig. 15). Because both the units and time scales are so different, direct comparison among these systems is like trying to compare apples and oranges. By non-dimensionalizing variables, one can derive ecologically important quantities that are comparable between these systems. Examples include the photosynthesis to respiration (P/R) ratio, the trophic transfer efficiency (fraction of biomass assimilated to next trophic level) and the coefficient of variation. These are examples of non-dimensional variables that allow us to meaningfully compare processes among ecosystems that may operate on different scales of time and space.

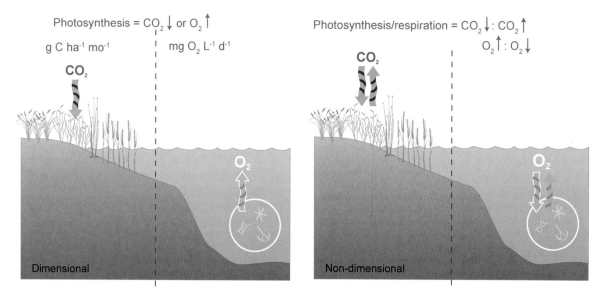

Figure 15: *Dimensional variables have units that often differ in different types of ecosystems. For example, for reasons associated with both method and time scale of development, photosynthesis might be measured in units of $g\,C\,ha^{-1}\,month^{-1}$ for marsh plants and in units of $mg\,O_2\,L^{-1}\,day^{-1}$ for plankton. Non-dimensional variables, such as the ratio of photosynthesis to respiration, do not have units and can therefore be more easily compared among ecosystems that differ in scale and assessment method.*

Dimensional analysis can be used to conserve key ecological relationships

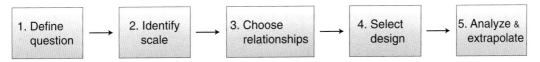

Figure 16: *There are five key steps to successfully applying a dimensional approach to the design of enclosed experimental systems (Petersen and Englund 2005): 1. Carefully define and refine the research question; 2. Identify the desired scale of inference and appropriate level of abstraction for the experimental system; 3. Choose key relationships that must be conserved and non-dimensional expressions that accomplish this; 4. Select an experimental design that conserves these relationships; 5. Analyze and extrapolate results to nature.*

As with other types of models, enclosed experimental ecosystems are simplifications and abstractions of nature. Selecting an appropriate level of abstraction for the system (Fig. 16, step 2) has direct bearing on how researchers achieve and appraise functional similarity and involves tradeoffs between control, realism, and scale. It is useful to distinguish between generic and ecosystem-specific models that represent the two extremes in this tradeoff (Chap. 2, Fig. 2). As the desired degree of specificity and desired level of realism increase, so too do the challenges and expense associated with selecting compensatory distortions to achieve dynamic similarity in the ecological relationships of interest.

Which of many ecological relationships to conserve (Fig. 16, step 3) depends on the research question and the ecology of the system. As a generalization, four overlapping environmental characteristics are key determinants of many ecological processes: habitat size, environmental variability, gradients, and interactions among habitats. (Fig. 17). In addition to being crucial features of most ecosystems, these are ecological attributes that are typically distorted in the construction of experimental ecosystems. Experimental artifacts, caused by the act of enclosing ecosystems, are an additional factor that may distort the dynamics of experimental ecosystems. The challenge is to develop a systematic dimensional approach that conserves functional similarity in key ecological relationships while minimizing the effects of experimental artifacts.

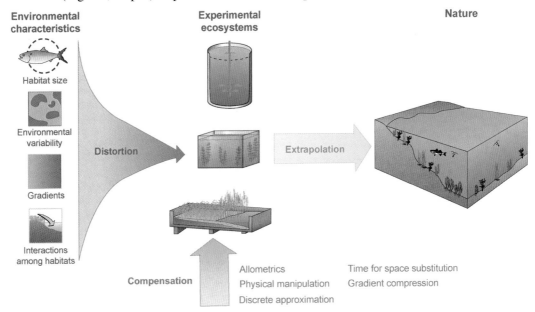

Figure 17: *Certain environmental characteristics are inevitably distorted in the construction of experimental ecosystems. A range of techniques and approaches (listed in green) can be used to compensate for these distortions in both experimental design and interpretation so that results can be systematically extrapolated to nature.*

The goal of the dimensional approach is to conserve effective scaling relationships

The goal of maintaining functional similarity between experimental and natural ecosystems requires that scaling attributes of organisms, processes, and the physical environment be carefully matched with each other. Often this amounts to preserving *effective scales* that are calculated by standardizing the absolute scaling attributes of the particular organisms or processes included in the experiment (for instance age, rate of movement, distance traveled) by their characteristic scales (e.g., generation time, maximum speed, home range) (Table 2). For example, the effective habitat size that an organism experiences (Crowley 1978) can be expressed as a function of the size, movement speed, generation times, and contact rates of the organisms involved relative to the size and physical heterogeneity of the environment that they occupy. Effective scales are dimensionless and can therefore be compared among systems that differ in absolute scale.

A property such as effective habitat size can be held the same in an experimental ecosystem as in nature by manipulating either the biological community employed in the experiment or the physical environment within the experimental ecosystem. The variety of approaches that are available for achieving this type of compensatory distortion are discussed on the following pages. The two central challenges of the dimensional approach lie in deciding what needs to be conserved among multiple ecological attributes, and in simultaneously conserving several quantities within the context of biological and experimental limitations.

Table 2: Examples that illustrate how different levels of organization have different absolute and characteristic scales. Effective scales (e.g., effective organism age) are calculated as functions of absolute and characteristic scales.

	Level of organization	Absolute scale	Characteristic scale
	Organism	Size, age, rate of movement, distance from point of reference	Adult size, life expectancy, generation time, maximum speed, home range
	Population	Number in group, current trophic level	Average number in group, average trophic position
	Community	Contact rate, predation rate	Maximum contact rate, maximum predation rate
	Ecosystem	Average age of dominant organisms, time since last major disturbance	Ecosystem size, lifespan of dominant organisms, reaction rates, frequency of disturbance, rate of recovery
	Example		

$$\text{Effective organism age} = \frac{f(\text{absolute scale})}{f(\text{characteristic scale})} = \frac{\text{Age}(y)}{\text{Life expectancy}(y)}$$

There are a number of techniques available for conserving ecological relationships

There are several general techniques available for conserving key ecological relationships (Table 3) including the following:

- Allometrics: This technique is used to relate a host of biological and ecological properties, such as lifespan, speed, and home range, to the size or body mass of organisms.

- Physical manipulation: Changes in the physical characteristics of the environment within the experimental ecosystem can sometimes be used to compensate for reduced scale.

- Time for space: Patchiness or spatial hetero-geneity can have an important effect on ecological processes, but may be impossible to directly incorporate in small enclosed experimental ecosystems. In some cases it is possible to simulate the ecological effects of spatial patchiness by subjecting the experimental ecosystems to changing inputs over time.

- Gradient compression: In nature, salinity, light, nutrient concentrations, organism abundance, and many other important features of the environment systematically vary with depth or with distance from a source. As with heterogeneity, experimental ecosystems may be too small to incorporate the gradient as it occurs in nature. There are, however, techniques for simulating the effects of the gradient on organisms and processes.

- Discrete approximation: In some circumstances multiple linked experimental ecosystems, with controlled degrees of exchange among them, may provide the most realistic approximation of the effects of spatial heterogeneity and gradients in nature.

These techniques can be applied alone or in combination and are discussed in detail on the following pages.

Table 3: Different techniques can be used to conserve different environmental attributes.

Attribute conserved	Techniques available				
	Allometrics	Physical manipulations	Time for space	Gradient compression	Discrete approximations
Habitat size	✓	✓	✗	✓	✓
Environmental variability	✗	✓	✓	✗	✓
Gradients	✗	✓	✓	✓	✓
Interactions among habitats	✗	✗	✗	✗	✓

Allometrics: Effective habitat size can be conserved by using small organisms

Figure 18: *General ecological relationships can often be explored in small experimental ecosystems by constructing communities composed of either very small organisms (e.g., bacteria and protozoa) or juvenile organisms (e.g., tree seedlings or saplings).*

Perhaps the most obvious option available for conserving relatively large effective scales at reduced absolute time and space scales is to construct mesocosm communities composed of small species or juvenile organisms (Fig. 18). The most extreme example of this may be 10^{-11} m³ microcapillary tubes that were successfully used as experimental systems to study spatial and temporal dynamics of competitive exclusion among protozoa feeding on herbivorous bacteria (Have 1990). More recently, this small-organism approach has received considerable attention as a means of elucidating general relationships between biological diversity and ecological function. (Lawton 1995; Naeem and Li 1998). These experiments take implicit advantage of the fact that the characteristic scales associated with organisms (i.e., lifespan, generation time, speed, home range, etc.) tend to increase with body size (Sheldon et al. 1972; Fig. 19). This approach is likely to be most useful for examining generic rather than ecosystem-specific research questions. For example, as Lawton (1995) wrote it is "unclear how easily results can be extrapolated between plants from very different functional groups."

General allometric equation

$$R = aM^b$$

R = Characteristic rate or spatial scale
M = Mass or size of organism
a, b = scaling coefficients

A relatively simple power-function is used to relate physiological and ecological properties of organisms to their body size. The particular value of scaling coefficients a and b tends to vary with the types of organisms or processes involved (e.g., metabolic activity as a function of body size for warm-blooded animals vs for cold-blooded animals).

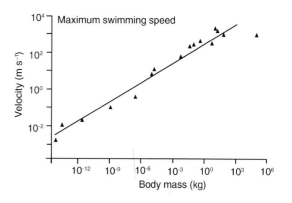

Figure 19: *The effect of organism size on swimming speed (Bonner 1965) maximum swim speed = 2.37*(body mass in kg)⁰·³⁵.*

Physical manipulation: Effective habitat size can be conserved by manipulating the physical environment

In addition to selecting communities of small organisms, a researcher can also potentially compensate for reduced space by altering the physical characteristics of the experimental ecosystem. For example, the longstanding problem of maintaining predator-prey coexistence in mesocosms without periodic re-inoculation (Gause 1934) has proven amenable to this approach.

One obvious physical manipulation that has been shown to increase the duration of predatory-prey coexistence is to simply increase mesocosm size and thereby reduce contact rates (Luckinbill 1974). A second option is to alter the physical environment. For example, Luckinbill was able to achieve predator-prey coexistence between protozoa in a physically homogenous environment by adding methyl cellulose (Fig. 20), which effectively increased the medium viscosity, thereby decreasing contact rates (Luckinbill 1973). A third option available for compensating for reduced absolute is to manipulate the spatial complexity of the environment.

Natural landscapes are often subdivided into patchy habitats with various degrees of isolation from each other. This type of spatial heterogeneity can be simulated under laboratory conditions by linking multiple individual container cells and allowing for varying degrees of exchange among the cells. This modular design is tessellated in the sense that each cell is akin to tile in the larger landscape mosaic. In recent years the tessellated approach has become an increasingly popular means of simulating generic landscape attributes in order to test metapopulation and community assembly theory. In many of these experiments, landscapes of mesocosms are assembled from experimentally defined communities of small organisms and housed in a number of identical microcosms, each representing a patch or an entire landscape. Studies of this type have explored the effects of manipulating a wide variety of landscape attributes including number of patches, patch size (grain), aggregate landscape size (extent), spatial configuration of patches, exchange between patches, and assembly sequence.

Huffaker (1958) performed a now classic set of experiments (Fig. 21) on predator-prey relations between two species of orange mites, in which landscapes of varying density and complexity were created out of oranges and rubber balls (Huffaker 1958). He found that spatial heterogeneity could, indeed, compensate for reduced size and lack of immigration, and his was the first experiment to successfully maintain multiple predator-prey oscillations in the absence of immigration.

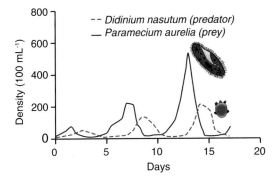

Figure 20: Addition of methyl cellulose slowed movement such that protozoa could exhibit classic predator-prey oscillation dynamics in a small microcosm (Luckinbill 1973).

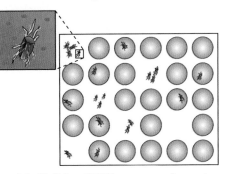

Figure 21: Huffaker (1958) constructed experimental landscapes composed of oranges and rubber balls that simulated spatial heterogeneity found in the natural environment such that predator and prey mites could coexist.

Time for space: Effective environmental variability can be conserved through time-for-space substitutions

As previously discussed, both the degree and the quality of spatial heterogeneity are recognized as important factors controlling dynamics in natural ecosystems (Hastings 1990; Duarte et al. 1992). Resources in nature are often patchy and the size of these patches may be much larger than the feasible size of the experimental ecosystem. Further, it is often desirable to generate a homogenous environment within a mesocosm (for instance by continuous mixing in a planktonic system) so that a single sample drawn at random can be used to characterize the whole system.

Time for space substitution conserves the effective environmental variability that organisms experience while simultaneously maintaining internal homogeneity. Time for space substitution means that the variability experienced by an organism as it moves through its habitat, be it predation pressure, nutrient levels, or light, is simulated in an experiment by subjecting organisms to temporal variation (Fig. 22). For example, in one experiment, continuous addition of nitrogen was used to simulate a spatially homogenous environment, leading to the dominance of a relatively small species of phytoplankton (Turpin and Harrison 1979). In contrast, a once-per-day nutrient addition simulated a patchy environment and favored a larger plankton species. Co-dominance occurred at an intermediate nutrient pulse frequency (Turpin and Harrison 1979).

In other experiments, time for space substitutions have been applied to examine the effects of variability in top-down control by planktivores. For instance, fish have been periodically added and removed from mesocosms to simulate the effects of schooling planktivores on phytoplankton. These experiments reveal that the top-down effects of grazing are more pronounced when fish are constantly present (press treatment) than they are when fish are added and removed (pulse treatment). As a final example, sinusoidal variation in light intensity has been used to simulate the variable light environment that phytoplankton experience as they move up and down through the water column (Gervais et al. 1999).

The so-called mean-field approximation is the assumption that homogeneous conditions within a mesocosm are a reasonable approximation of conditions that are more heterogeneous in nature. The results of theory and of the research just described suggest that experiments designed with this assumption may lead to distorted ecological dynamics.

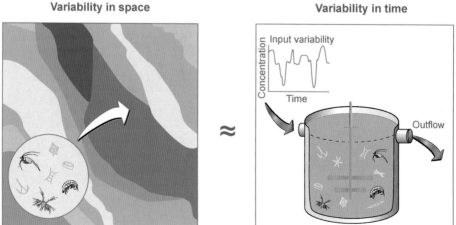

Figure 22: Movement of plankton through an environment with localized regions of high nutrients (pictured as different shades of blue) can be simulated in a mesocosm by varying the frequency and magnitude of input.

Gradient compression: A deep-water light environment can be simulated in a shallow mesocosm

Biological, chemical, and physical gradients are both a cause and an effect of ecological interactions. For instance, light energy is absorbed by plant and non-living material as it passes down through a marsh canopy or water column, and this has important effects on ecosystem energetics. Likewise, species composition and ecological processes vary with the salinity gradient moving longitudinally from the head to the mouth of an estuary, and crosswise with depth moving from littoral to pelagic zones. The reduced length scales inherent to mesocosms constrain the space available for interactions that occur in response to many ecologically important gradients (Fig. 23). Several dimensional approaches can be taken to compensate for reduced length scale in order to maintain dynamic similarity with respect to natural gradients.

Primary productivity and other processes in pelagic ecosystems are strongly influenced by the light gradient that organisms experience as they move up and down through the water column. Light generally exhibits an exponential decay in intensity with depth, as described by the Beer-Lambert law in Eq. (1):

$$I_z = I_o e^{-kdz} \qquad (1)$$

where I_o = surface light intensity (e.g., μmols m^{-2} s^{-1}), I_z = light intensity at depth z, k_d = the light attenuation coefficient (m^{-1}), and z = depth (m). When z is taken as the depth of the mixed layer (or depth of the water column in shallow systems), the product $k_d z$ is referred to as optical depth and provides a good description of the vertical light environment experienced by phytoplankton.

The fact that optical depth is a non-dimensional variable suggests an option for counterbalancing the reduced depth of most aquatic mesocosms. That is, reductions in depth might be compensated for by increasing the light attenuation coefficient. The specific target value of k_d necessary to achieve functional similarity in the light environment can be determined with a scaling factor; light attenuation in the mesocosms (k_{dM}) should be proportional to attenuation in the natural ecosystem (k_{dN}) adjusted for the decrease in depth. This light scaling factor can be expressed as in equation (2):

$$k_{dM} = k_{dN} \left(\frac{z_M}{z_N} \right) \qquad (2)$$

where z_N and z_M are depth in nature and in the mesocosm respectively.

Reports suggest that k_d might be controlled by manipulating the optical properties of container walls and tank dimensions (Fig. 56 and Fig. 57). Specifically, k_d appears to increase with light absorption by walls and with decreasing tank radius (Kemp unpublished data; Nixon et al. 1980; Peeters et al. 1993). The addition of different concentrations of biologically inert dye also provides a means to control light attenuation in order to simulate the optical depth of a deep water column in a shallow mesocosm (Fig. 24, see next page).

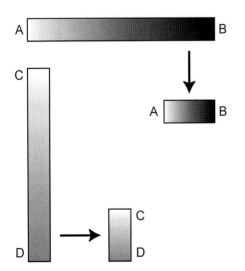

Figure 23: Horizontal and vertical gradients in salinity, light, nutrients, and other ecologically important factors are typically eliminated or compressed in experimental ecosystems.

Gradient compression: A deep-water light environment can be simulated in a shallow mesocosm

As discussed on the preceding page, conserving optical depth ensures that phytoplankton experience the same range of light intensities from surface waters to the bottom of the tank that they experience in a deep water column in nature. However, there are other important features of the light environment, such as the rate of change (fluctuations) in light intensity, that phytoplankton experience as they move up and down through the mixed layer. Fluctuations that they experience in light intensity determine the physiological responses phytoplankton will exhibit (Lewis et al. 1984). Unfortunately, if the vertical mixing environment (k_z = vertical eddy diffusivity, Table 2) is held the same in a mesocosm as in nature, then conservation of optical depth will necessarily ensure that the phytoplankton experience more rapid fluctuations in light intensity in the mesocosm than in the deeper natural system. Thus, simultaneous conservation of both light variability and the light gradient entails compensatory distortion in mixing as well as light attenuation. The key variable to conserve in order to achieve dynamic similarity in light fluctuation is the mixing time (T_m), which is a measure of the average time it takes for a particle to circulate up and down through the water column equation (1). Mixing time can be expressed as a function of k_z:

$$T_m = \frac{z^2}{2k_z} \qquad (1)$$

where z = depth of the mixed layer or mesocosm. Setting mixing time equal in nature and the mesocosm and solving for eddy diffusivity equation one finds an appropriately scaled k_z of:

$$k_{zM} = k_{zN} \left(\frac{z_M}{z_N} \right)^2 \qquad (2)$$

Thus, to maintain similar light environments for phytoplankton in mesocosms with different depths, a researcher could intentionally manipulate both light attenuation and eddy diffusivity as illustrated in Figure 24.

The example on these last two pages focuses on achieving similarity in the light environment. However, literature reviews indicate that it is increasingly feasible to quantify and to manipulate the physical environment in experimental ecosystems to achieve functional similarity in a broad range of important environmental characteristics (Sanford 1997; Petersen and Hastings 2001).

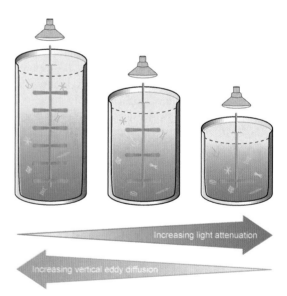

Figure 24: *The light gradient that phytoplankton experience as they move up and down through a deep water column can be simulated in a relatively shallow mesocosm by manipulations that increase the light attenuation coefficient k_d. Conserving the same variability in light intensity for the phytoplankton also requires that vertical mixing (eddy diffusion coefficient, k_z) be reduced in the shallow mesocosm relative to the deep water column being simulated.*

Gradient compression: Horizontal gradients can also be condensed to conserve realism

Gradients are ubiquitous in nature and important for many ecological processes. Fick's law, which describes molecular diffusion in response to a concentration gradient, is essentially identical in form to Darcy's law, which describes the flow of water through a porous medium in response to a pressure gradient. As with the light gradient described in previous pages, the distances over which these gradients occur in nature are often greater than the feasible size of enclosed experimental ecosystems (Fig. 25). Compensatory distortions provide a means of conserving key features of the gradient.

Consider a situation in which one wishes to design an experimental estuarine marsh ecosystem that retains both the groundwater residence time and tidal amplitude of a much longer natural marsh, equation 1. Darcy's law governs flow through the marsh substrate:

$$Q = k_p \frac{\Delta H}{L} \qquad (1)$$

where Q = flow rate per unit cross-sectional area (m s^{-1}), k_p = the hydraulic conductivity of the marsh media (m s^{-1}), ΔH = tidal amplitude (m), and L = length of marsh (Fig. 25). Residence time can be defined as $T_r = L/Q$, and if one re-expresses

Darcy's law in terms of residence time equation 2 it becomes:

$$T_r = k_p \frac{L^2}{\Delta H} . \qquad (2)$$

Our objective is to construct an experimental marsh that has a reduced length but that retains the same residence time and tidal amplitude as the natural system. One can create separate equations for the natural marsh and mesocosm and designate the hydraulic conductivity of the natural reference marsh as k_{pN} and mesocosm as k_{pM} (equation 3). Because an equivalent residence time is desired, one can set the equations equal to each other and solve for k_{pM}:

$$k_{pM} = k_{pN} \left(\frac{L_M}{L_N} \right)^2 . \qquad (3)$$

Hydraulic conductivity of the mesocosm k_{pM} could be manipulated by altering the porosity of the sediment medium. A modification of this variable may generate unacceptable distortions in sediment biogeochemistry, but at a minimum the dimensional approach makes the tradeoffs involved in design decisions explicit rather than implicit.

Figure 25: Groundwater residence time and tidal amplitude in a marsh mesocosm (top panel) can be made to simulate a larger natural marsh (bottom panel) by increasing slope and decreasing the porosity of the substrate in the mesocosm.

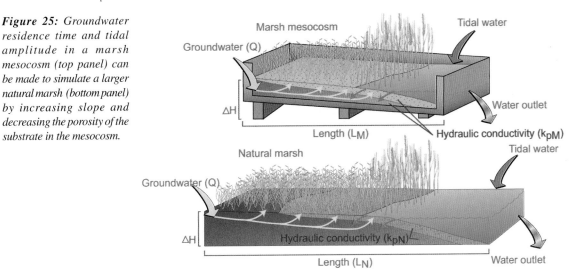

Discrete approximation: Interactions among habitats can be conserved with multicosms

The examples presented in the preceding pages conserve effective gradients by means of compensatory distortions that compress relatively large-scale gradients that occur in nature within much smaller experimental systems (Fig. 26). In many cases the gradient of interest may be too large to compress within a single mesocosm, or the condensed gradient may bring organisms or biogeochemical processes into artificially close proximity, thereby inducing unrealistic dynamics. An alternate approach is to simulate the gradient with a series of linked modular mesocosms (multicosms).

Figure 26: MEERC marsh mesocosms. These mesocosms are designed to simulate environmental conditions in a single habitat.

Mesocosms are typically designed to simulate environmental conditions within a single habitat type. However, in nature the biotic and abiotic interactions among functionally different habitat types are often as important as strictly internal interactions in determining overall ecological dynamics. An option for studying these among-system interactions is to control exchange among a series of discrete, coupled, mesocosms (Fig. 27). This multicosm approach is similar to both the tessellated mesocosms and the estuarine gradient discussed on preceeding pages in that reductions in space are counterbalanced by means of controlled flow among compartments. However, in multicosms the properties of individual cells are functionally distinct. For instance, mesocosm cells may represent a series of distinct adjacent habitats (e.g., upland → marsh → littoral → open water), or interacting subsystems segregated by functional group (e.g., producer, consumer, decomposer, or predator and prey).

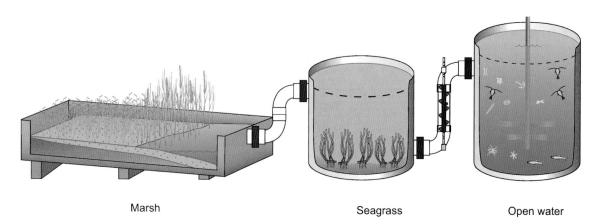

Marsh   Seagrass   Open water

Figure 27: Multicosms physically link habitats to simulate controlled exchange and study multiple habitats simultaneously. For example, an Archemedes screw pump links a wetlands mesocosm with a seagrass mesocosm and that mesocosm is linked with a planktonic mesocosm (p. 79).

Discrete approximation: Continuous gradients in nature can be simulated with discrete compartments in experimental ecosystems

Another alternative to maintaining functional similarity is to hold conditions within each compartment relatively constant, but to alter conditions among compartments in accordance with position along a gradient. This approach is analogous to the finite difference approach to solving differential equations; continuous change is approximated with discrete, incremental change (Fig. 28).

This modular approach to gradients can be used, for instance, to solve the light gradient problem. At the extreme, a light gradient can be represented with just two compartments located at the endpoints of the gradient, in this case a fully lighted zone and a dark zone (King 1980). A more sophisticated apparatus for simulating light fields has been generated in which acrylic tubes are passed down through a series of black rubber sheets that segregate the water column into discrete zones, each of which is independently illuminated from the side (see lettered compartments in Fig. 29; Estrada et al. 1987).

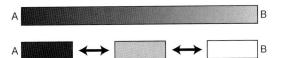

Figure 28: Continuous gradients that occur over broad scales in nature (top panel) can be simulated by exchanging material among discrete mesocosm compartments, each of which represents a distinct zone in the gradient (bottom panel).

Perhaps the most widely applied substitution of discrete for continuous gradients has occurred in models of the estuarine salinity gradient (Margalef 1967; Doering et al. 1995). Some estuarine models have been so elaborate as to include special devices that allow large organisms, such as crabs, to migrate up and down the salinity gradient (Adey and Loveland 1998). In many cases, particularly for ecosystem-specific models, this discrete approach may provide a more realistic model of nature than is possible with the continuous approaches to compensatory distortions described in the preceding pages.

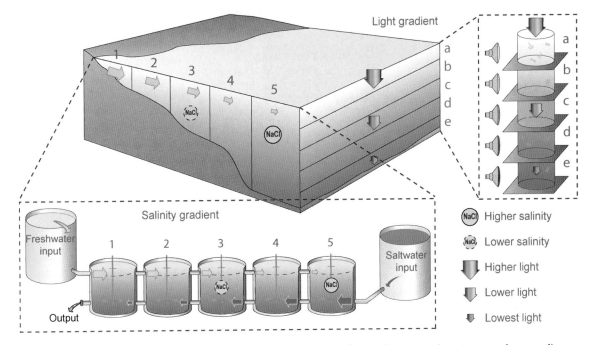

Figure 29: Salinity and light gradients that are continuous in nature can be simulated by exchanging water between discrete compartments in mesocosms.

Compensatory distortions can achieve dynamic similarity

Ecological relationships that are important in nature can often be maintained in experimental ecosystems by means of experimental designs that intentionally compensate for inherent reductions in scale (Table 4; Petersen and Hastings 2001).

Table 4: *Examples of ecological properties, experimental approaches, and ecologial relationship conserved by compensatory distortions (Petersen and Hastings 2001).*

Ecological property	Experimental approach	Ecological relationship conserved
Effective time and space scales	• Manipulate organism size: choose communities composed of small organisms for mesocosms. • Manipulate physical environment: change viscosity of medium or accentuate spatial heterogeneity.	• Characteristic time and space scales of organisms relative to size of habitat. • Available habitat relative to rate of organism movement.
Effective environmental variability	• Create tessellated multicosm landscapes; set up separate or semi-separate mesocosms with controlled exchange of organisms among units.	• Exchange among patches.
Effective gradients	• Condense a continuous vertical gradient; for example, manipulate physical environment so as to alter light attenuation and mixing coefficients. • Condense a continuous horizontal gradient; alter hydraulic conductivity by manipulating the porosity of the sediment. • Condense a large continuous gradient; use discrete, zoned compartments.	• Mixing time for particles to circulate though the water column, and light gradient. • Residence time and tidal amplitude. • Light gradient. • Salinity gradient.
Interactions among habitats	• Use multicosms; control exchange among a series of fundamentally different mesocosms habitat.	• Interactions, exchanges, and feedbacks among different habitats.

Dimensional approaches can be used to compensate for artifacts of scale in experimental design

Regardless of how successfully one is in designing experimental systems that conserve natural relationships, one is still faced with scaling artifacts of enclosure. Although a distinct issue from the types of compensatory distortions discussed on preceding pages, dimensional considerations can also potentially be used to minimize these artifacts and to maximize comparability among experiments that differ in scale. As an example, it has been experimentally demonstrated that the relative ecological effect of the undesirable periphytic community living attached both to the walls of experimental aquatic mesocosms is related to the radius of the mesocosm and to the interval at which the walls are cleaned (Fig. 65; Chen and Kemp 2004; Chen et al. 1997). Parallel experiments are often conducted in different sized experimental systems. In order to facilitate comparisons among systems that differ in size, it might be desirable to select a cleaning regime that maintains an equivalent level of wall growth in the different sized systems (Fig. 30). For instance, in a system with wall periphyton and water column phytoplankton, one might select a cleaning regime that maintains total wall biomass at time t (W_t) below 10% of the combined phytoplankton (P_t) and wall biomass:

$$\frac{W_t}{P_t + W_t} < 0.1. \tag{1}$$

A variety of formulations might be appropriate for estimating the respective growth rates of wall periphyton and phytoplankton, equation 1. A simple approach would be to assume that wall periphyton grows exponentially per unit wall area such that total wall growth at time t is:

$$W_t = (W_o e^{kt})(2\pi r z) \tag{2}$$

where r and z are tank radius and depth respectively ($2\pi r z$ = wall surface area), W_o is initial periphyton biomass expressed per unit wall area, k = the growth rate of wall periphyton,

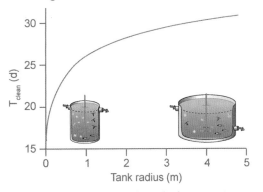

Figure 30: *The relative effect of edge organisms on ecological dynamics decreases with increasing mesocosm radius in non-resuspension systems. The relative contribution of periphyton that grow on mesocosm walls to total biomass can be held below 10% by cleaning the walls at an interval (T_{clean}) that is determined by mesocosm dimension.*

and t=time. As a simple (and not entirely unrealistic) scenario, one can assume that total producer biomass ($P + W$) is constrained by available resources to some fixed level, in which case total phytoplankton biomass can simply be expressed as:

$$P = P_o(\pi r^2 z) - W \tag{3}$$

where P_o = initial phytoplankton biomass per unit volume. Substituting Eqs. (2) and (3) into Eq. (1), and solving for time:

$$T_{clean} = \frac{\ln(r) + \ln(0.05 \times P_o / W_o)}{k} \tag{4}$$

where T_{clean} = the time that it takes for wall biomass to achieve 10% of total system biomass in different radius systems (Fig. 30). For instance, in mesocosms of 1 m and 10 m radius, where P_o = 100 mg C/m³, W_o = 1 mg C/m², and k = 0.5 day⁻¹, equation 4 indicates cleaning intervals of 3 and 8 days respectively. Although equation 4 has restricted applicability, the example illustrates that it is possible to systematically scale experimental design in order to maintain artifactual effects within acceptable levels and at equivalent levels in differently scaled systems and non-resuspension systems.

Dimensional approaches allow comparisons of results among ecosystems and experiments, although there are limitations

The examples assembled in this section illustrate that a diverse variety of dimensional approaches are available for designing experimental ecosystems so as to conserve key ecological relationships. Even if dimensional approaches are not used to design experiments, careful consideration of dimensions can often be used to meaningfully compare results among different experiments. For example, one might wish to compare the developmental patterns of primary producers in marsh and planktonic ecosystems (Fig. 15). Because producers have very different time scales for reproduction, developmental patterns cannot be compared directly. However, they can be meaningfully compared if time series of plant biomass in experimental plankton and marsh ecosystems are graphed using non-dimensional time and biomass (Fig. 31).

Limitations in dimensional approaches associated with biological constraints roughly fall into four categories (Petersen and Hastings 2001): *inflexibility*, *specificity*, *equivalency*, and *interdependence*. Limitations associated with the *inflexibility* and *specificity* of biological variables and relationships are particularly pronounced in ecosystem-specific models where organism size, generation time, and relationships among organism are constrained by the particular ecosystem under investigation, leaving manipulation of the physical environment as the only option for achieving functional similarity. The issue of *equivalency* is of special concern for generic models used to elucidate general ecological principles that apply to many different kinds of ecological systems. Whether there really are general principles relating variables such as ecological complexity to ecological function remains an open question that will not be easily resolved with even the cleverest of dimensional approaches. *Interdependence* among biological variables is problematic for both ecosystem-specific and generic mesocosm experiments in that it ensures that compensatory distortions that increase realism for one organism or set of relationships typically also decrease realism for others.

The dimensional compromises that will maximize realism in an experiment are a function of the ecological processes and relationships associated with the research question. Collectively, the limitations described above make it impossible to develop a single recipe for applying dimensional approaches to experimental design. Nevertheless, the approach and examples outlined here provide a useful framework for designing experimental ecosystems that use dimensional considerations to retain key functional relationships present in larger natural ecosystems.

Figure 31: Patterns of plant community development in marsh and planktonic ecosystems occur on very different time scales, but are similar when data are expressed as non-dimensional variables. Effective time scale used on the x-axis = (time elapsed in experiment)/(characteristic turnover time for primary producers). Turnover times used were 1.4 day for phytoplankton and 3 month for marsh plants. Biomass on the y-axis is graphed as a percentage of maximum biomass recorded in each experiment ($4 kg m^{-2}$ dry weight for marsh plants and $14 \mu g L^{-1}$ chlorophyll a for phytoplankton; Petersen and Englund 2005).

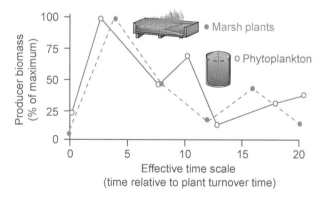

Modeling tools
J.E. Petersen and W.M. Kemp

Modeling allows for the manipulation of time and space scales beyond the bounds of experimental ecosystems

Models can synthesize and integrate findings from experiments conducted using the full spectrum of empirical research approaches (Fig. 32, top panel). Three key steps that can be distinguished in the simulation modeling process are *formulation*, *calibration*, and *validation*. Different scales of research play different roles in the modeling process. For example, laboratory experiments are used to estimate coefficient values, such as the rate constants that are incorporated into the model as it is formulated. Field and laboratory experiments can then be used to calibrate the model. Calibration entails modifying a model so that it realistically simulates observed dynamics. Validation involves comparing calibrated model outputs with data independently collected in mesocosms or observed in natural systems. Models calibrated and validated with empirically collected data can be used to predict responses at scales and in response to ecological conditions that cannot be easily duplicated in either experimental ecosystems or in nature (Fig. 32, bottom panel).

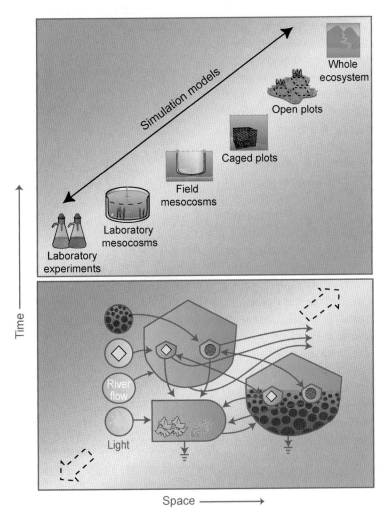

Figure 32: *Models can help researchers synthesize and interpret data collected using different experimental approaches (top panel). In particular, simulation models can predict responses to different spatial and temporal scales (bottom panel).*

Simulation models are valuable tools for designing and interpreting experiments

Dynamic simulation models and mesocosms are abstractions and simplifications of nature, representing different balances among *realism*, *control*, and *generality* (Kemp et al. 1980). One can think of numerical and living models as nested hierarchical structures, in which smaller subsystems are contained within larger systems (Fig. 33).

Time and space scales are easier to manipulate in models than in mesocosms, whereas realism (the degree to which experimental results represent the target natural ecosystem) is higher in mesocosms than in simulation models because mesocosms include properties of adaptation and self-organization not easily represented in the computer code of simulation models. Conversely, control (the ability to relate cause and effect, to replicate and to repeat experiments) is greater in mesocosms than in nature, and greater yet in simulation models than in mesocosms. Finally, mesocosm experiments usually require that control and generality (the breadth of different systems to which results are applicable) be compromised to allow a mesocosm to realistically match the target natural ecosystem (Nixon et al. 1980), whereas simulation model formulations can be adjusted to enhance general applicability to a broad range of ecosystems and conditions. Some experimentalists have chosen to enhance realism in their studies by increasing the size and complexity of experimental systems and the duration of studies, all at the expense of replication and control (Carpenter 1996; Schindler 1998). In any case, the fundamental complementarity of simulation models and mesocosms is partially reflected in the frequency with which mesocosm studies and simulation models have been integrated into individual ecosystem studies (Brockmann 1990; Parsons 1990; Baretta-Bekker et al. 1994).

Simulation models can assist in design, analysis, and extrapolation of experimental ecosystem research, improving the efficacy of such research (Fig. 34; Hill and Wiegert 1980;

Brinkman et al. 1994). Mesocosm studies offer an excellent opportunity for developing coefficient values and quantifying relationships that can then be used to develop well-calibrated simulation models (Nixon et al. 1979). In fact, rich data streams from mesocosm investigations can support sophisticated simulation model calibration methods (Vallino 2000). On the other hand, discrepancies between predictions and observations in parallel research involving both numerical models and experimental ecosystems can, in principle, reveal limitations in the mechanistic understanding of the dynamics of experimental or natural ecosystems (Beyers and Odum 1993).

Figure 33: Simulation models can help synthesize data from mesocosms conducted at different scales such as this tethered mesocosm being deployed from a boat into a natural system.

Simulation models

Living models

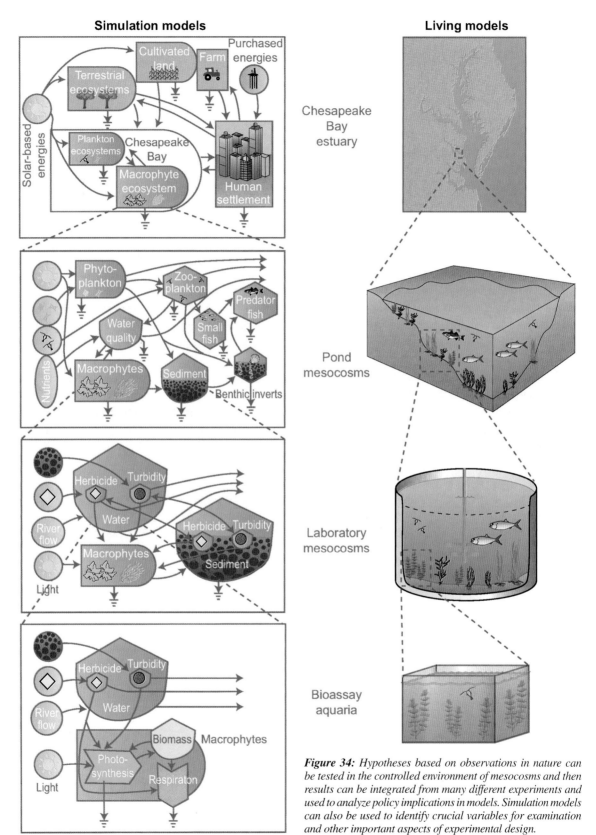

Figure 34: Hypotheses based on observations in nature can be tested in the controlled environment of mesocosms and then results can be integrated from many different experiments and used to analyze policy implications in models. Simulation models can also be used to identify crucial variables for examination and other important aspects of experimental design.

Simulation models can be used to improve the design of mesocosm facilities and experiments

Simulation models are useful during the experimental design process for identifying essential physical factors that must be recreated in mesocosms, key variables that should be measured, and appropriate sampling protocols.

For example, a preliminary model of salt marsh mesocosms containing *Spartina alterniflora* was used to decide on appropriate plant sampling protocols within MEERC (Zelenke and Madden 1996). Model simulations indicated that frequent collection of plant biomass samples could seriously disrupt ecosystem dynamics, but that reduced sampling frequency was sufficient to capture essential responses without inducing disruption (Fig. 35). Similarly, initial simulation studies of experimental aquatic grass ecosystems were used to select the number and size of herbivorous grazers used in manipulative mesocosm studies (Bartleson et al. 2005).

The outcomes of these efforts demonstrate that simulation models can play a valuable role in

---

Simulation models can identify the following:

- Essential physical factors that should be accurately recreated;

- Key variables that should be measured; and

- Appropriate sampling protocols for experimental design.

---

improving experimental design. A caveat is that, in order to serve this important function, simulation models obviously need to be functional (though not necessarily calibrated or very final) before or in association with the physical construction of the experimental ecosystems (Brinkman et al. 1994). In many prior studies, modeling efforts often took place after experiments were completed, which prevented them from serving this critical function.

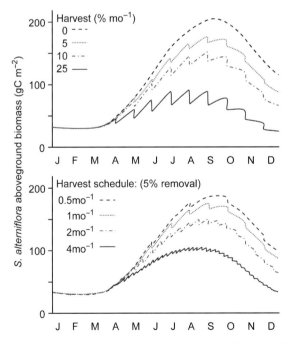

Figure 35: *Model response to alternative protocols for sampling aboveground biomass of* Spartina alterniflora *for variable sample sizes (upper panel) and sampling frequencies (lower panel).*

Simulation models can be used to interpret and extrapolate results from mesocosms to natural ecosystems

Simplifications and modifications in scale make it difficult to directly extrapolate the results of experimental ecosystems to nature. In additional to dimensional analysis discussed earlier in this section, simulation models provide a tool for incorporating the effects of simplifications and distortions in scale so that results can be meaningfully interpreted and extrapolated. Earlier studies (Nixon et al. 1979; Kemp et al. 1980; Brinkman et al. 1994) suggested that simulation models carefully calibrated with data from mesocosm perturbation studies are effective devices for predicting responses of natural ecosystems to similar perturbations.

A model that was initially developed to aid in the design of mesocosm experiments can be calibrated with experimental results and then be used to enhance interpretation and improve the design of subsequent experiments. For instance, in the MEERC program a simulation model of experimental ecosystems containing aquatic grasses was calibrated and used to better understand the interacting effects of nutrients and herbivorous grazing on community dynamics, including growth of epiphytic algae and their host vascular plants (Bartleson et al. 2005). In this case, simulations led researchers to the understanding that aquatic grass growth was more responsive to grazing under lower nutrient conditions than under high nutrient conditions. This was because higher nutrient loading rates tended to overwhelm the ability of herbivores to remove epiphytic material and these epiphytes shaded aquatic grass leaves (Fig. 36). Similarly, a simple simulation model of plankton dynamics (Petersen et al. 2003) effectively explained what appeared to be inconsistent phytoplankton and mesozooplankton responses to variations in mesocosm water exchange rates. Here,

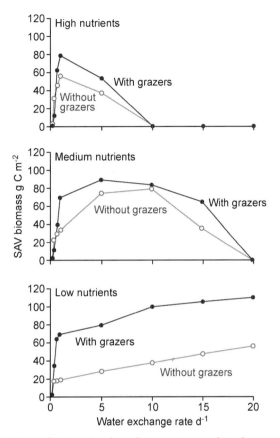

Figure 36: *Graphs of simulation output resulting from a model calibrated with the results of mesocosm experiments. These graphs show the effects of variable nutrient loading rates on the biomass of submerged aquatic vegetation (SAV) with and without the addition of epiphytic grazers (adapted from Bartleson et al. 2005).*

simulation experiments demonstrated that the nature of plankton responses to changes in water exchange rate depends on external nutrient concentrations; increased water exchange appears to elicit bottom-up control with low external nutrients, but top-down control with relatively high external nutrient levels.

Fundamental effects of scale and artifacts of enclosure should be represented in simulation models of mesocosms

Simulation models allow for inclusion of a range of important processes and factors that, for practical reasons, must be excluded in mesocosm experiments. Among other things, simulation models provide an ideal opportunity to incorporate and account for scaling differences between experimental ecosystems and the natural ecosystems that they are designed to represent. A problem is that until recently there were insufficient data available with which to calibrate models to account for the effects of both artifacts of enclosure and fundamental effects of scale. Findings from the multiscale experiments described in this book have been used to incorporate scaling relationships into simulation model formulations. This enables scale-sensitive extrapolations from smaller experimental scales to the larger scales of nature.

For example, a model developed to simulate dynamic interactions between coastal planktonic, benthic, and periphytic communities was calibrated and tested with data from experiments conducted in different sizes and shapes of experimental ecosystems (Fig. 37; Chen 1998). Simulation model experiments provided quantitative predictions of how total gross primary production (GPP expressed per unit volume) is affected by water column depth under different light and nutrient regimes. Since depth is a fundamental scaling attribute in nature, these results allow us to build models in which data can be extrapolated from relatively shallow mesocosms to deeper natural waters.

Calibrating this same simulation model with data on periphyton growth allowed researchers to quantify artifacts of scale associated with the community of organisms that inhabit the walls of mesocosms.

Simulation results suggest that, under the conditions explored in these mesocosm experiments, artifacts associated with periphyton growth on mesocosm walls could be reduced

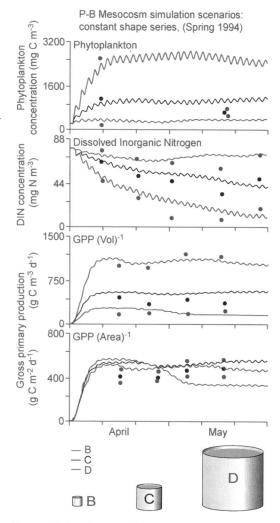

Figure 37: Simulation model output (solid lines) and data from experiments (round dots) for experiments conducted in experimental ecosystems of different dimensions under high nutrient conditions.The different mesocosms were identical in shape (ratio of radius to depth), but differed in size, depth, and surface areas.

to acceptable levels with twice-weekly wall cleaning or by using tanks with a radius of 2 m or greater. Simulations with no-wall scenarios were then used to remove the effect of this artifact and more realistically explore planktonic-benthic interactions in nature.

Simulation models of mesocosms allow the exchanges that occur in heterogeneous landscapes to be examined

Coastal ecosystems are characterized by heterogeneous spatial distributions of organisms, resources, habitats, and processes. Highly variable exchange rates of biological material and energy are critical characteristics of these ecosystems. Components of exchange that are excluded from experimental ecosystems can be considered in simulation models. However, while simulation models that consider conditions within a relatively homogenous body of water are valuable, they may not be sufficient to capture dynamics that result from heterogeneity. Whereas mesocosm studies and simple simulation models provide well-controlled measurements of responses to experimental manipulations within relatively homogeneous systems, understanding how responses to perturbations would differ with variable material exchange among linked patchy water masses (Fig. 38) requires alternative approaches. One alternative is to physically link experimental ecosystems that contain different habitats. A second approach, spatially explicit numerical models, can examine the effects of water exchange among heterogeneous environments.

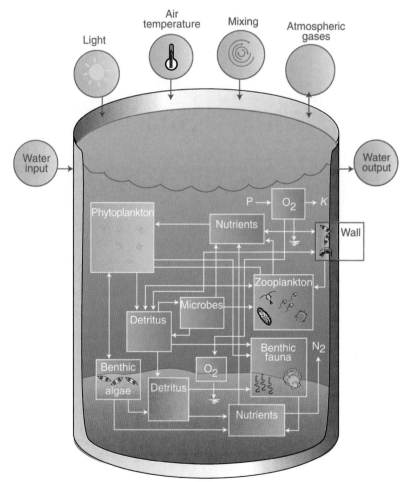

Figure 38: A unit model of exchanges and internal processes controlling the ecological dynamics within a pelagic-benthic mesocosm. Models such as this are directly useful for examining dynamics within single habitats. Spatially explicit models couple multiple unit models of different habitats to explore landscape-scale dynamics resulting from interactions among habitats.

Spatially explicit simulation models provide a higher-level tool for integrating, synthesizing, and extrapolating results of experiments conducted in different habitats

Given the importance of spatial heterogeneity in controlling ecological dynamics, MEERC researchers believe that coupling mesocosms with spatially explicitly dynamic simulation models will become an increasingly powerful approach to ecological research. In this approach, mesocosms can be thought of as individual cells (grain) within a heterogeneous matrix of different habitats that cover broad spatial extent (Fig. 39).

Likewise, models can be used to explore effects of temporal variability that are difficult to incorporate in the design of mesocosm studies. Scale effects are often non-linear and may sometimes exhibit threshold effects. Numerical models offer an excellent tool for exploring non-linear feedback effects at scales that are larger than individual mesocosms and involve interactions among habitats.

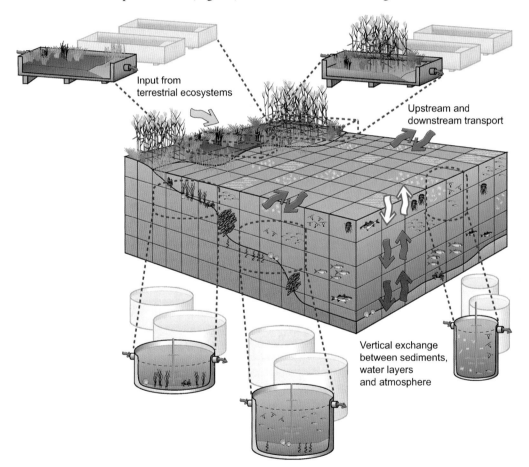

Figure 39: *The estuary can be conceptualized as a matrix. Each grid cell represents a particular set of ecological conditions that collectively form a heterogeneous landscape (or seascape). In mesocosm research, isolated cells represent different habitats from the larger landscape to provide experimental control and replication. In order to realistically simulate landscape-scale interactions, researchers need to consider biological, material, and energetic exchanges among these cells. Mesocosm research is therefore complemented by spatially explicit approaches to mathematical or numerical modeling that allow for the simulation of large areas over long time scales with complex patterns of exchange. For models, mesocosms, and nature, scaling dimensions (i.e., size, shape, exchange rates, and ecological complexity within each grid cell or patch) can have profound effects on the ecological dynamics observed.*

References

Adey, W.H. and K. Loveland. 1998. Dynamic Aquaria: Building Living Ecosystems, 2nd edition. Academic Press, San Diego, CA.

Baretta-Bekker, J.G., B. Riemann, J.W. Baretta and E. Koch Rasmussen. 1994. Testing the microbial loop concept by comparing mesocosm data with results from a dynamical simulation model. Marine Ecology-Progress Series 106:187–198.

Bartleson, R.D., W.M. Kemp and J.C. Stevenson. 2005. Use of a simulation model to examine effects of nutrient loading and grazing on *Potamogeton perfoliatus L.* communities in microcosms. Ecological Modeling 185:483–512.

Berg, G.M., P.M. Glibert and C.-C. Chen. 1999. Dimension effects of enclosures on ecological processes in pelagic systems. Limnology and Oceanography 44:1331–1340.

Beyers, R.J. and H.T. Odum. 1993. Ecological Microcosms. Springer-Verlag, NY.

Bonner, J.T. 1965. Size and Cycle: An Essay on the Structure of Biology. Princeton University Press, Princeton, NJ.

Brinkman, A.G., C.J.M. Philippart and G. Holtrop. 1994. Mesocosms and ecosystem modeling. Vie et Milieu 44:29–37.

Brockmann, U. 1990. Pelagic mesocosms: II. Process studies. in C.M. Lalli (ed.). Enclosed Experimental Marine Ecosystems: A Review and Recommendations. p. 81–108. Springer-Verlag, NY.

Carpenter, S.R. 1996. Microcosm experiments have limited relevance for community and ecosystem ecology. Ecology 77:667–680.

Chen, C.-C. 1998. Wall effects in estuarine mesocosms: Scaling experiments and simulation models. PhD dissertation. University of Maryland, College Park, MD.

Chen, C.-C. and W.M. Kemp. 2004. Periphyton communities in experimental marine ecosystems: Scaling the effects of removal from container walls. Marine Ecology Progress Series. 271:27–41.

Chen, C.-C., J.E. Petersen and W.M. Kemp. 1997. Spatial and temporal scaling of periphyton growth on walls of estuarine mesocosms. Marine Ecology-Progress Series 155:1–15.

Crowder, M.J. and D.J. Hand. 1990. Analysis of Repeated Measures. Chapman and Hall, London.

Crowley, P.H. 1978. Effective size and the persistence of ecosystems. Oecologia 35:185–195.

Doering, P.H., C.A. Oviatt, B.L. Nowicki, E.G. Klos and L.W. Reed. 1995. Phosphorus and nitrogen limitation of primary production in a simulated estuarine gradient. Marine Ecology-Progress Series 124:271–287.

Duarte, C.M., M. Masó and M. Merino. 1992. The relationship between mesoscale phytoplankton heterogeneity and hydrographic variability. Deep-Sea Research 39:45–54.

Estrada, M., M. Alcaraz and C. Marrasé. 1987. Effect of reversed light gradients on the phytoplankton composition in marine microcosms. Investigación Pesquera 51:443–458.

Gause, G.F. 1934. The Struggle for Existence. Williams and Wilkins, Baltimore, MD.

Gervais, F., T. Hintze and H. Behrendt. 1999. An incubator for the simulation of a fluctuating light climate in studies of planktonic primary productivity. International Review of Hydrobiology 84:49–60.

Hastings, A. 1990. Spatial heterogeneity and ecological models. Ecology 71:426–428.

Have, A. 1990. Microslides as microcosms for the study of ciliate communities. Transactions of the American Microscopical Society 109:129–140.

Hill, J. and R.G. Wiegert. 1980. Microcosms in ecological modeling. Pages 138–163 in J.P. Giesy, Jr. (ed.). Microcosms in Ecological Research. National Technical Information Service, Springfield, VA.

Huffaker, C.B. 1958. Experimental studies on predation: Dispersion factors and predator-prey oscillations. Hilgardia 27:343–383.

Kemp, W.M., M.R. Lewis, J.J. Cunningham, J.C. Stevenson and W.R. Boynton. 1980. Microcosms, macrophytes, and hierarchies: Environmental research in the Chesapeake Bay. Pages 911–936 in J.P. Giesy, Jr. (ed.). Microcosms in Ecological Research. National Technical Information Service, Springfield, VA.

King, D.L. 1980. Some cautions in applying results from aquatic microcosms. Pages 164–191 in J.P. Giesy, Jr. (ed.). Microcosms in Ecological Research. National Technical Information Service, Springfield, VA.

Lawton, J.H. 1995. Ecological experiments with model systems. Science 269:328–331.

Lewis, M.R., J.J. Cullen and T. Platt. 1984. Relationships between vertical mixing and photoadaptation of phytoplankton: Similarity criteria. Marine Ecology-Progress Series 15:141–149.

Luckinbill, L.S. 1973. Coexistence in laboratory populations of *Paramecium aurelia* and its predator *Didinium nasutum*. Ecology 54:1320–1327.

Luckinbill, L.S. 1974. The effects of space and enrichment on a predator-prey system. Ecology 55:1142–1147.

Margalef, R. 1967. Laboratory analogues of estuarine plankton systems. Pages 515–521 in G. Lauff (ed.). Estuaries. American Association for the Advancement of Science, Washington D.C.

Naeem, S. and S. Li. 1998. Consumer species richness and autotrophic biomass. Ecology 79:2603–2615.

Nixon, S.W., D. Alonso, M.E.Q. Pilson and B.A. Buckley. 1980. Turbulent mixing in aquatic mesocosms. Pages 818–849 in J.P. Giesy, Jr. (ed.). Microcosms in Ecological Research. National Technical Information Service, Springfield, VA.

Nixon, S.W., C.A. Oviatt, J.N. Kremer and K. Perez. 1979. The use of numerical models and laboratory microcosms in estuarine ecosystem analysis – simulations of a winter phytoplankton bloom. Pages 165–188 in R.F. Dame (ed.). Marsh-Estuarine Systems Simulation. University of South Carolina Press, Columbia, SC.

Pandey, G., S. Lovejoy and D. Schertzer. 1998. Multifractal analysis of daily river flows including extremes for basins of five to two million square kilometres, one day to 75 years. Journal of Hydrology 208:62–81.

Parsons, T.R. 1990. The use of mathematical models in conjunction with mesocosm ecosystem research in C.M. Lalli (ed.). Enclosed Experimental Marine Ecosystems: A Review and Recommendations. p. 197–210. Springer-Verlag, NY.

Peeters, J.C.H., F. Arts, V. Escaravage, H.A. Haas, J.E.A. de Jong, R. van Loon, B. Moest and A. van der Put. 1993. Studies on light climate, mixing and reproducibility of ecosystem variables in mesocosms: Consequences for the design. Pages 7–23 in J.C.H. Peeters, J.C.A. Joordens, A.C. Smaal and P.H. Nienhuis (eds.). The Impact of Marine

Eutrophication on Phytoplankton and Benthic Suspension Feeders: Results of a Mesocosm Pilot Study. Report No. DGW-93.039, NIOO-CEMO-654, Middelburg, Netherlands.

Petersen, J.E., C.-C. Chen and W.M. Kemp. 1997. Scaling aquatic primary productivity: Experiments under nutrient- and light-limited conditions. Ecology 78:2326–2338.

Petersen, J.E., J.C. Cornwell and W.M. Kemp. 1999. Implicit scaling in the design of experimental aquatic ecosystems. Oikos 85:3–18.

Petersen, J.E. andG. Englund. 2005. Dimensional approaches to designing better experimental ecosystems: A practitioners guide with examples. Oecologia 145:216–224.

Petersen, J.E. and A. Hastings. 2001. Dimensional approaches to scaling experimental ecosystems: Designing mousetraps to catch elephants. American Naturalist 157:324–333.

Petersen, J.E., W.M. Kemp, R. Bartleson, W.R. Boynton, C.-C. Chen, J.C. Cornwell, R.H. Gardner, D.C. Hinkle, E.D. Houde, T.C. Malone, W.P. Mowitt, L. Murray, L.P. Sanford, J.C. Stevenson, K.L. Sundberg and S.E. Suttles. 2003. Multiscale experiments in coastal ecology: Improving realism and advancing theory. BioScience 53:1181–1197.

Platt, T. and K.L. Denman. 1975. Spectral analysis in ecology. Annual Review of Ecology and Systematics 6:189–210.

Sanford, L.P. 1997. Turbulent mixing in experimental ecosystem studies. Marine Ecology-Progress Series 161:265–293.

Schindler, D.W. 1998. Replication versus realism: The need for ecosystemscale experiments. Ecosystems 1:323–334.

Sheldon, R.W., A. Prakash and W.H. Sutcliffe Jr,. 1972. The size distribution of particles in the ocean. Limnology and Oceanography 17:327–340.

Tilman, D. 1989. Ecological experimentation: Strengths and conceptual problems. Pages 136–157 in G.E. Likens (ed.). Long-Term Studies in Ecology. Springer-Verlag, NY.

Turpin, D.H. and P.J. Harrison. 1979. Limiting nutrient patchiness and its role in phytoplankton ecology. Journal of Experimental Marine Biology and Ecology 39:151–166.

Vallino, J.J. 2000. Improving marine ecosystem models: Use of data assimilation and mesocosm experiments. Journal of Marine Research 58:117–164.

Zelenke, J. and C.J., Madden. 1996. Simulation model of biogeochemical processes in marsh mesocosms Report to U.S. EPA, Annapolis, MD.

Management Applications

Scientific results from enclosed experimental ecosystems can be particularly useful for management applications. Mesocosm results have the advantage of being controlled, replicated experiments that simulate essential features of nature. Furthermore, mesocosms can be used to develop explicit scaling relationships between laboratory and field observations, thereby enhancing quantitative interpretation of results from scientific experiments conducted in laboratory or field settings. By being intermediate in spatial, temporal, and complexity scales between strictly controlled laboratory experiments and field observations, mesocosm results can provide context for interpreting laboratory and field results.

Mesocosms can explore ecosystem responses to human perturbation and result in information relevant to management applications. This information can lead to paradigm shifts in the way scientists approach environmental problems. For example, the MERL tanks at the University of Rhode Island were being replenished with Narragansett Bay water at the onset of a harmful algal bloom (brown tide; *Aureococcus anophagefferens*) and only the control tanks without nutrient addition contained dense concentrations of the brown tide organisms. This observation from mesocosms directed future research into the origin and cause of this harmful algal bloom in fundamental ways even without explicit brown tide experiments conducted in the mesocosms (Cosper et al. 1987).

Coastal ecosystems are under increasing threats globally, and providing scientific results to better understand and manage coastal issues is increasingly important. Experimental ecosystems are tools scientists and resource managers can use to provide relevant results at the time, space, and complexity scales necessary for management decisions. While mesocosm research can lead to increased understanding of fundamental processes in ecology, it is the management applications that are often the underlying rationale for mesocosm research.

In this section, six different management application case studies are elucidated with (a) a problem description delineating the management issue, (b) a brief summary of relevant research findings, and (c) a discussion of the management implications. There are far more applications that can be developed from mesocosm results. However, rather than providing an exhaustive list of applications, selected case studies are used to illustrate the diversity of management issues that can be addressed by using mesocosm results. In addition, the case studies illustrate how to link relevant mesocosm research to management recommendations.

Nutrient enrichment and aquatic grass restoration
W.M. Kemp and L. Murray

Submerged aquatic grasses have experienced declines globally, in large part due to eutrophication stimulating phytoplankton and epiphyte growth which compete with aquatic grasses for light. Plant habitat requirements for light and nutrients have been crucial in developing water quality criteria and restoration goals for aquatic grasses in Chesapeake Bay. Using mesocosm experiments, simulation models, field transplants, and scaling concepts, the dynamics of aquatic grass meadows in absorbing nutrients and reducing epiphyte shading were investigated. Aquatic grass meadows that are relatively intact provide an important nursery function for transplanted grasses due to their nutrient absorptive and water clarification capacities, and field studies demonstrate the relative success of transplanting native grasses into existing meadows. These results aid in optimizing aquatic grass restoration efforts on degraded meadows and increasing the conservation value of existing meadows.

Problem description

Over the last several decades, populations of submerged aquatic grass have experienced dramatic declines in abundance at coastal sites throughout the world (Short and Wylie-Echeverria 1996). Although the causes of these losses in plant abundance are unclear in some cases, most appear to be related to parallel trends of increasing nutrient and sediment inputs from surrounding watersheds to coastal environments (Daurte 1995). Foremost among the many ecological consequences of these declines is the loss of productive habitat for crabs, lobsters, fish, waterfowl, and marine mammals (Kemp 2000). Submerged aquatic grass beds also serve to clarify the water column by reducing tidal currents and surface waves, thereby trapping suspended particles and stabilizing bottom sediments (Gacia and Duarte 2001). In addition, these beds tend to stimulate rates of nutrient cycling and to act as nutrient sinks (Nagel 2007). Substantial research effort has been directed toward understanding the dynamics of these submerged grass communities in relation to environmental conditions.

Large-scale losses of submerged aquatic grass in Chesapeake Bay, which were first noted in the late 1960s, have continued through the present day (Kemp et al. 2006). Throughout this large estuarine ecosystem, more than a dozen aquatic grass species have experienced declines in abundance, with initial effects centered in the upper Bay and later losses observed in lower Bay regions. An example (Fig. 1) of the magnitude of this aquatic grass decline is evident in a pair of aerial photographs

Figure 1: *Aerial photographs of shallow estuarine waters surrounding Solomons Island, MD (USA) taken in summer in the 1950s (left) and in 1979 (right). Extensive aquatic grass beds (dark areas) and clear water around the island are apparent in the earlier photo, while more turbid water and absence of grass beds are evident in the later photo (Stankelis et al. 2003).*

of Solomons Island at the mouth of the Patuxent River (Stankelis et al. 2003). Field observations and mesocosm experiments revealed that aquatic grass declines were associated with increased concentrations of nutrients that stimulate growth of planktonic and epiphytic (growing on plant leaves) algae (Kemp et al. 1983; Twilley et al. 1985). Phytoplankton and other suspended particles block light as it passes down through the water to the canopy, while epiphytic algae further shade these plants at the leaf surface (Fig. 2). Subsequent studies in other coastal ecosystems confirmed the relationship between nutrient enrichment and reduction of light required for growth of aquatic grasses (Borum 1985; Short et al. 1995; Tomasko et al. 1996). Climate variations have modified these trends. For example, a massive flooding event in 1972 led to precipitous declines in upper Bay aquatic grasses while a drought (1999–2002) contributed to plant resurgence in the middle Bay (Kemp et al. 2005). In both cases, nutrient and

sediment loading from diffuse watershed sources were directly proportional to rainfall.

State and federal resource management agencies have committed to reducing nutrient and sediment loading to the Bay and to initiating aquatic grass restoration activities. However, questions have lingered regarding how results from mesocosm experiments should be extrapolated to conditions in nature and how field observations on water quality and aquatic grasses can be related to nutrient and sediment loading rates.

Research findings

Research in MEERC and related studies has helped to address these questions by using mesocosm experiments (Sturgis and Murray 1997) and field observations (Schulte 2003) to develop models and scaling relationships that simulate aquatic grass responses to and interactions with light and nutrient environmental conditions (Madden and Kemp 1996; Bartleson

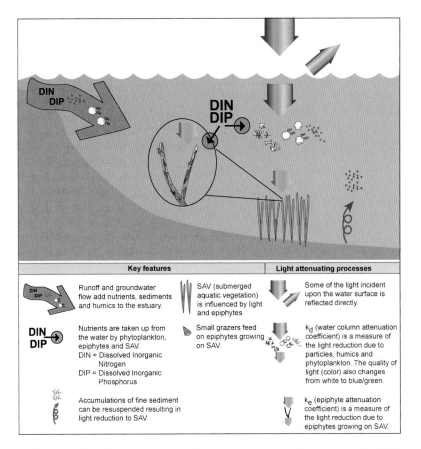

Figure 2: Conceptual diagram of how light attenuation affects aquatic grasses (Dennison et al. 1993).

et al. 2005). Model functions were developed and calibrated from data, and then transferred to management models for assessing nutrient and sediment management alternatives (Cerco and Moore 2001). In this sense, the simulation models were used as tools to scale-up from experimental observations to conditions in the natural Bay. These same calibrated model equations were also used to develop water quality criteria for assessing success of management efforts by relating monitoring data to habitat conditions needed for aquatic grass survival (Kemp et al. 2004). Further MEERC research has applied results of modeling and field studies to develop novel approaches for restoring aquatic grass species into Bay habitats where they formerly flourished (Melton 2002; Hengst 2007).

A prerequisite for the restoration of aquatic grass beds in Chesapeake Bay is the improvement in water quality conditions, particularly nutrients, total suspended solids (TSS), algal biomass (chlorophyll a), and water clarity. Since 1985, the Chesapeake Bay water quality monitoring program has been providing measurements (at 140 stations and 2–4 week intervals) of key properties to track progress in Bay restoration. Initial

efforts to use simple statistical analysis of these data for predicting aquatic grass presence were instructive but ambiguous (Dennsion et al. 1993). For example, water clarity measures were used to compute underwater light conditions, and nutrient levels were assumed to serve as indices of epiphyte growth and associated shading on plant leaves. However, epiphyte effects on light available to aquatic grasses were not quantitatively estimated. Using a combination of mesocosm data and simulation model equations, MEERC researchers developed algorithms to calculate epiphyte shading from monitoring data. In this calculation, epiphyte biomass is computed from nutrient and light conditions based on output from a model calibrated with mesocosm data (Fig. 3). Shading effects of epiphyte biomass are then computed in terms of reduction of light reaching plant leaves. Bay-wide comparisons of aquatic grass habitat suitability computed from this algorithm versus actual plant distributions agreed remarkably well (Kemp et al. 2004; Madden and Kemp 1996).

Although growth of aquatic plants is regulated by nutrient concentrations and light availability in the overlying water, modeling studies supported by

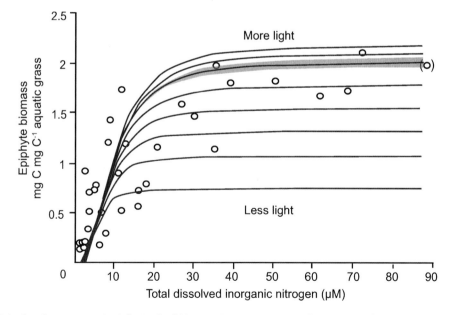

Figure 3: Calculated responses of epiphytic algal biomass (per aquatic grass biomass) to changes in dissolved inorganic nitrogen (DIN) concentration under various light conditions in estuarine waters. Circles are data obtained from mesocosm experiments conducted at a fixed light level with varying DIN concentrations. A simulation model was calibrated with this data (Sturgis and Murray 1997) (gray line third from top) and then a family of curves was developed by manipulation light intensity in the simulation model (Kemp et al. 2004).

MEERC have shown that the plant beds themselves modify nutrient and light conditions in their local environment. Healthy plant beds inhibit growth of epiphytic algae by assimilating nutrients from the water column, and they increase water clarity by trapping and depositing algal chlorophyll *a* and TSS from the water column to the bed sediments. The ability of aquatic plants to modify these environmental conditions is dependent in part on the rate that water is exchanged between the bed and surrounding areas. Model experiments demonstrated the tendency of plant beds to reduce TSS concentrations and associated turbidity in overlying water (Cerco and Moore 2001). Further simulations revealed that epiphyte growth on aquatic grass leaves is regulated by interactions between input nutrient concentrations, plant biomass density, and water exchange rate (Fig. 4). Although aquatic grasses are able to out-compete epiphytic algae for nutrients in larger, denser beds, epiphyte growth tends to overcome this effect with higher nutrient concentrations and exchange rates

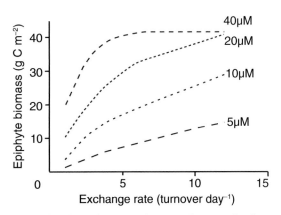

Figure 4: Effects of water exchange and nutrient loading rate (per aquatic grass biomass) on biomass of epiphytic algae (after Bartleson et al. 2005).

(Bartleson et al. 2005). Comparative analysis of aquatic grass beds of different size, density, and shape (patchiness) results in scaling relationships confirming that epiphyte abundance was inversely related to bed density, (percent cover) and positively related to bed patchiness (Fig. 5). Plants

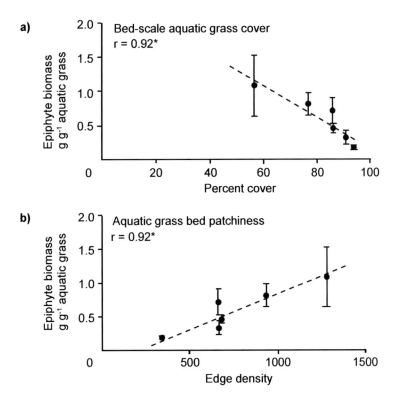

Figure 5: Comparative analysis of R. maritima *beds illustrating the effects of (***a***) plant cover and (***b***) edge density (patchiness) on growth of epiphytic algae on plant leaves (Schutle 2003).*

Figure 6: Experimental transplanting of Potamogeton perfoliatus *into the mesohaline regions of Chesapeake Bay involves raising vegetative cuttings in healthly plant-sediment plugs in a green house (**a**,**b**) and transferring these plants into natural habitats (**c**,**d**).*

growing in denser, more continuous beds tended to retard growth of epiphytic algae by efficiently absorbing nutrients from the water (Schulte 2003). This information allows managers to assess the health of submerged aquatic plants by interpreting annual aerial photographic surveys in terms of bed density, size, and patchiness. It also suggests the existence of thresholds of aquatic

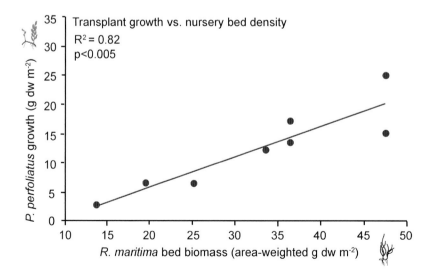

Figure 7: Relationship between growth of transplanted Potamogeton perfoliatus *and density of nursery beds of* Ruppia maritima *for restoration sites in the Choptank River estuarine system (Hengst 2007).*

plant bed density, size, and patchiness beyond which beds will be unable to sustain themselves. Current spin-off research is further examining this question.

As noted above, although the mesohaline portion of Chesapeake Bay once supported a diversity of submerged aquatic grass species, most of these disappeared from former habitats in the early 1970s. MEERC studies have focused on developing efficient methods for restoring native aquatic grass species back to this Bay region (Fig. 6). During the last 15 years, one pioneer aquatic grass species, *Ruppia maritima* has exhibited modest and variable recovery in this Bay region, with strongest growth in drought years (1992, 1995, and 1999–2002). Recent research emerging from MEERC has shown that these stands of *R. maritima* can be used as nursery beds that enhance local nutrient and water-clarity conditions sufficiently to promote growth of other aquatic grass species transplanted into these beds. To be effective, the scale of these nursery beds needs to be sufficient to improve substantially water quality conditions for growth of transplants. For example, growth of redhead grass (*Potamogeton perfoliatus*) transplanted into open areas within these nursery

beds was shown to be directly related to the density of the surrounding *R. maritima* stands (Hengst 2007) (Fig. 7). Subsequent field surveys have revealed that transplanted plots of *P. perfoliatus* and a related species (*Stuckenia pectinata*) had self-propagated within 2–4 years to form multiple satellite colonies arising within a 300 m radius from these transplanted plots. These satellite colonies increased the effective area of restored beds by three to five fold (Fig. 8). Although the detailed mechanisms whereby these satellite colonies were formed are unclear, evidence suggests that fragmented stem sections bearing leaves and roots are readily transported to and rooted within nearby open sediments.

Management implications
This MEERC research has contributed in several important ways to the development of management strategies for improving estuarine environmental conditions to support restoration of submerged aquatic grasses. It has helped to develop habitat requirements for growth and survival of submersed aquatic grasses in the estuary, and these requirements have, in turn, contributed to the development of water quality criteria (light requirements) that serve as goals for developing total maximum daily loads

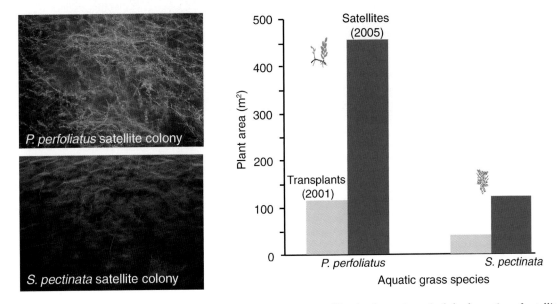

Figure 8: Self-propagation of transplanted Potamogeton perfoliatus *and* Stuckenia pectinata *beds by formation of satellite patches within 4 years in areas surrounded by nursery beds of* Ruppia maritima. *Bar chart shows expansion from original transplanted areas. Photos show large patches of the transplanted species growing up in* R. maritima *beds.*

(TMDL) for nutrients and sediments. The aquatic grass light requirements provide a tool for using annual water quality monitoring data to assess trends in Bay recovery. The algorithms developed in this research to relate environmental conditions to aquatic grass growth and survival have been incorporated into a large-scale coupled model of estuarine circulation and ecological dynamics (Cerco and Moore 2001) for use in generating alternative water quality management scenarios (Fig. 9).

In a sense, these models serve as quantitative tools for extrapolating results from mesocosm studies to the scales of the whole Bay, where they can be used to address nutrient and sediment management questions. Field research has shown that the spatial scales of aquatic plant distribution (bed size, density, and patchiness) are critical characteristics that reflect the health of the plant beds and their effectiveness in modifying environmental conditions at local and regional scales. This information allows managers to assess the response of submerged aquatic grass beds to water quality management programs and climate variations. Finally, MEERC-related research in mesocosm experiments, field studies, and simulation modeling analyses has contributed to the development of effective protocols that can be used by managers for rapid restoration of stable multi-species submersed plant communities whereby pioneer stands of *R. maritima* serve as nursery beds.

Impacted ecosystem | Unimpacted ecosystem

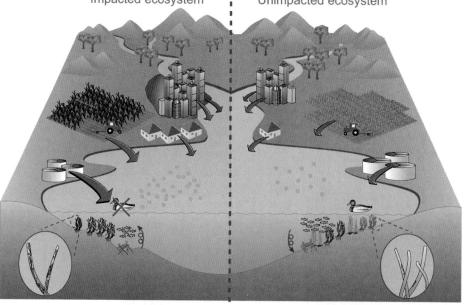

 Increased nutrient and sediment loads from urbanization, sewage discharge, and agriculture contribute to aquatic grass loss.

 Large pulses of nutrient inputs promote epiphytes and inhibits seagrass growth.

Unhealthy aquatic grass beds cannot serve as nurseries for fish and shellfish; thus, impacting commerical fisheries.

 Where aquatic grass is not present to stabilize sediment, resuspension is common.

 Reduced nutrient and sediment loads from smart growth, sewage treatment upgrades, and conservation agriculture practices such as cover crops promote healthy aquatic grasses.

 Smaller pluses of nutrient inputs result in fewer epiphytes and more seagrass growth.

 Healthy aquatic grass beds provide nursery functions for fish, crabs, and lobsters.

Aquatic grass beds provide sediment stabilization and improve water clarity.

Figure 9: *Aquatic grass beds serve many ecosytems functions including nursery habitat, sediment stabilization, and water clarity.*

Tidal marshes, nutrient filtration, and rising sea level
J.C. Stevenson

Tidal marshes that line the shores of many estuaries effectively absorb nutrients, providing a key ecosystem service in areas with excess nutrient enrichment (Fig. 10). The nutrient absorptive capacity of tidal marshes with high and low tidal marsh biodiversity (plants and consumers) was tested in mesocosm experiments. Results from these multi-year mesocosm studies revealed little effect of biodiversity—low diversity marshes were essentially equivalent to high diversity marshes in removing nutrients, providing important guidance for large scale marsh restoration activities in Chesapeake Bay. The effect of relative sea level rise due to subsidence and rising sea levels on tidal marshes is also a major concern, and mesocosm results provided support for augmenting tidal

marshes with application of dredged sediments and nutrient enrichments from agricultural runoff.

Problem description

Despite overall nutrent budgets of various wetlands, showing that about ½ of the incoming nitrate in groundwater can be removed by marshes, little has actually been quantified concerning how much groundwater is actually processed by fringe marshes surrounding Chesapeake Bay. Thus, these have been consistently left out of previous nutrient budgets of the Chesapeake Bay.

After 5 years of development, the experimental marshes, which simulated a 6 m width of fringe marsh in nature, were able to buffer

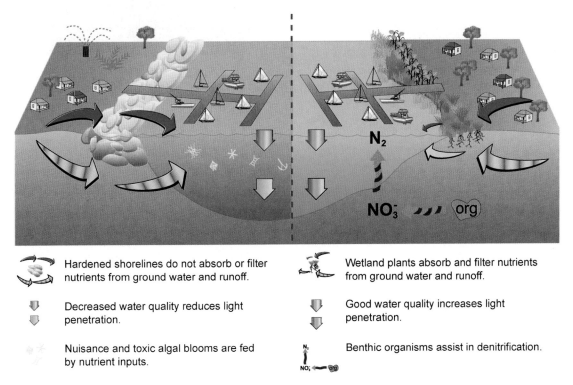

Hardened shorelines do not absorb or filter nutrients from ground water and runoff.

Wetland plants absorb and filter nutrients from ground water and runoff.

Decreased water quality reduces light penetration.

Good water quality increases light penetration.

Nuisance and toxic algal blooms are fed by nutrient inputs.

Benthic organisms assist in denitrification.

Figure 10: *Tidal marshes provide many ecosystem services including nutrient absorption and filtration and improved water quality.*

upwards of 70% of the incoming nitrate at about 10 mg L^{-1}. Furthermore, a comparison of denitrification in the summer using N_2:Ar ratios showed that the relatively newly-constructed mesocosms actually have lower rates than nearby natural marshes.

The studies show that fringe marshes of Chesapeake Bay are very important in buffering high nitrate in groundwater inputs (e.g., from agriculture). However, since many of these marshes are rapidly eroding (particularly as the decadal rise of sea-level was 1.2 cm per year from 1989–1999 at Baltimore) their ability to perform that service in the Chesapeake Bay is declining. Thus, the nutrient "clean-up" in terms of non-point sources will be even more difficult in the 21st century. This argues strongly that more serious attention has to be paid, not only to restore fringe marshes where possible, but to reducing agricultural inputs (by using 100% cover crops, reducing farm acreage, etc.) in areas where erosion (e.g. southern Dorchester County, western Somerset County, etc.) is severe, in order to achieve water quality goals for the Chesapeake Bay and its tributaries. There is both local and global concern that lowering species diversity of many ecosystems (Fig. 12) may diminish their productivity and limit their functional ability to provide services in the future. For example, lowered biodiversity particularly lowers gross

Figure 12: Low div ersity marsh (top) and high diversity marsh (bottom).

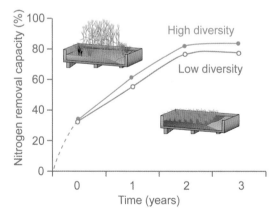

Figure 11: Figure of groundwater nutrient uptake in mesocosms for 1996–2000.

primary production in small terrestrial ecosystems by lowering respiration rates (Naeem et al. 1994; Naeem et al. 1996). This has led to concern that restoration projects need to diversify plantings in order to maximize functional equivalence to ecosystems damaged or lost because of various anthropogenic activities. However no diversity studies using replication have been done on tidal marsh systems.

After 5 years of data collected in MEERC research it was concluded that initial planting of six plant species with three consumers (versus two plant species and no consumers) did not change overall rates of net production or nutrient buffering capacity in the mesocosms (Fig. 11).

Research findings

Diversifying tidal marsh plantings with numerous (sometimes poorly adapted) species appears to be a waste of resources. It is better to match the species with the geochemical conditions in the sediments and not spend excessive amounts of money and time on attempting to diversify artificially created marshes. This information should be taken into account for marsh restoration. At Poplar Island initial restoration involving seeding of over 20 species was attempted but had little success. Instead, plant plugs should be selected based on the salinity, redox, and nutrient regime in the receiving sediments.

There is a need to dispose of an excess of over 2–3 million cubic meters of dredged materials in Chesapeake Bay every year for the foreseeable future to accommodate maintenance of the approaches to Baltimore Harbor. A portion of this material was used to build new marshes at Poplar Island and other locations, in part to make up for the losses in tidal marsh due to rising sea-level (Fig. 13). There are continuing questions about how fast marshes can actually be established on dredged material, which is generally high in phosphorus (P), nitrogen (N) and sometimes contains pyrite. This is especially

pertinent since the first test marsh cell at Poplar Island was mostly composed of local sand rather than channel material.

Two sizes of marsh mesocosms were established during the course of MEERC research. In the 3 m-long mesocosms, relatively coarse grained material from a fringe marsh was used and whole sections of turf were planted over it with great success. Experiments with fine-grained substances were also conducted. In the 6 m-long mesocosms, hand-dredged, fine-grained high-P materials from Lakes Cove at Horn Point Laboratory was added. Then, 7.5 cm inch cores containing marsh grasses were extracted. In the August of 1994, these cores were planted (10 cm apart) in the mesocosm in a regularly spaced grid. In spite of the fact that there were high levels of sulfide in 1994, the next year plant growth was phenomenally high (mostly due to high available P in sediments) with peak biomass reaching over $5\,kg\,m^2$ in some mesocosms.

Management implications

These studies demonstrate that fine-grained dredged materials can support luxuriant *Spartina alterniflora* marshes which have very high above-ground biomass. In addition, less salinity

Figure 13: *Excess sediments from the navigation channels in Baltimore Harbor have been used to rebuild the marshes lost on Poplar Island. Marsh erosion and loss was caused by sea-level rise.*

tolerant species such as Olney's Threesquare (*Schoenoplectus americanus*), were established in similar fashion. Because there was rapid rhizomatous growth in the months after the plugs were placed in the substrate, weedy species such as *Phragmites australis* did not invade these systems. Only on one occasion did *P. australis* actually invade our mesocosms, in an area where the substrate was left bare after sediment cores were removed from the marsh. Thus, even though there were large *Phragmites* stands nearby supplying ample wind borne seeds, invasion did not occur where the marsh plants established healthy stands within the first year.

At the outset of the MEERC studies there was concern that agricultural outputs might be accelerating marsh peat degradation (Fig.14) at places like Blackwater National Wildlife Refuge where over 5,000 hectares have been converted from marsh to open water. Nitrate is a potent oxidant and when present could enhance conversion of peat to CO_2. Although there is presently not much agriculture around Blackwater National Wildlife Refuge, 25 years ago it was ringed with farms and this was the period when marsh loss was greatest.

When high-nitrogen groundwater was introduced to the marsh mesocosms it lowered redox considerably compared to controls. However, this resulted in more plant biomass (below and above ground) and increased overall peat production.

High N inputs actually appear beneficial to surrounding marshes. The MEERC marsh studies strongly suggested that agriculture could be very helpful in terms of providing N-rich groundwater to natural and restored marshes. Restored salt marsh grasses near San Diego did not have the necessary stature (30 stems per meter > 90 cm tall) to support desired clapper rail populations (*Rallus longirostris levipes*) in one of the restoration projects at Sweetwater National Wildlife Refuge (Zedler and Callaway 1999). This marsh creation project had little or no groundwater and this probably resulted in low redox and stunted plants (especially in a low rainfall, high temperature climate). Installation of dispersal systems to introduce high N in the subsurface (e.g., from a nearby wastewater treatment plant), would undoubtedly alleviate the problem and help reduce excess sewage.

Eroding marshes   **Stabilized marshes**

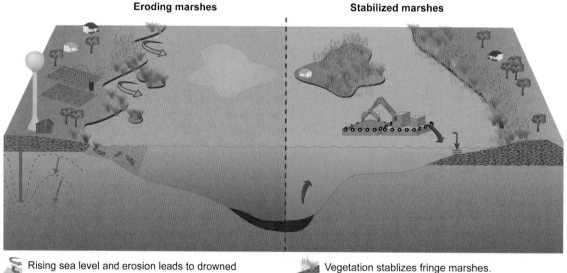

 Rising sea level and erosion leads to drowned fringe marshes.

Intense ground water use lowers the water table.

Vegetation stablizes fringe marshes.

Addition of fine-grain dredge material supports marsh plants.

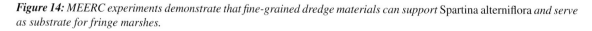

Figure 14: MEERC experiments demonstrate that fine-grained dredge materials can support Spartina alterniflora *and serve as substrate for fringe marshes.*

Nitrogen and phytoplankton blooms
P.M. Glibert and G.M. Berg

Excess nutrient inputs often stimulate phytoplankton growth leading to eutrophication and algal blooms, including harmful algal blooms. Experiments with controlled doses of nitrogen allowed for scaling effects of mesocosm to be factored into the understanding of how the fate of nitrogen may vary under different conditions.

Problem description

Phytoplankton growth is dependent on many factors, among which light, temperature, and nutrients are central. These are factors that are easily manipulated within mesocosms; mesocosms further allow the effects of these variables on phytoplankton to be studied in the broader context of the microbial and trophic interactions that occur in a natural, or semi-natural community (Fig. 15). While mesocosms are extremely useful in studies of energy and material transfer between plankton and benthic communities, artifacts and constraints of enclosure pose limitations on how experimental results can be extrapolated to natural systems. Many of these constraints and artifacts have already been described in the preceding chapters. Regardless of scaling relationships between enclosed systems and their natural counterparts, the development and responses of communities of organisms within mesocosm systems will never precisely mimic those of the natural environment (Fig. 16). Yet, the understanding of the constraints of enclosure can be insightful in terms of understanding specific variables that impact specific pathways of nutrient cycling and nutrient use by phytoplankton.

Research findings

Mesocosm scale can drive pathways of nitrogen cycling. The processes of primary productivity and nutrient uptake are dependent on light and cellular energy. In natural systems, light attenuation is a function of the phytoplankton itself, dissolved constituents such as humic material, and water

*Figure 15: (**a**) Mesocoms allow the effects of light, temperature, and nutrients on phytoplankton to be studied. Though the conditions in mesocosm do not exactly mimic natural systems (**b**), mesocosms can help researchers understand microbial nutrient cycling such as may occur during algal blooms.*

(Kirk 1994). In mesocosms, on the other hand, the vertical attenuation of light is also a function of mesocosm geometry. In the MEERC mesocosms, total light energy showed more than a twofold range among the different mesocosm types. Dimensions A and B received the most light energy, C and E received less and D received the least. Light attenuation was a result of mesocosm radius and the distance of the top between the tank and the light source.

Does light energy or availability matter to nitrogen cycling? In MEERC mesocosms, the change in light attenuation with depth was closely related to the fate of nitrate. Nitrate is typically the dominant form of nitrogen in coastal and estuarine systems, and thus understanding its

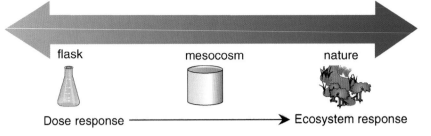

Figure 16: *Communities of organisms in mesocoms and natural systems will respond differently to variables and inputs. However, mesocosm can provide insight into energy and material transfer between plankton and benthic organisms.*

fate is important to algal ecology, and also to ecological modeling and management. In the MEERC mesocosms, large differences in nitrate fluxes were found among the various mesocosm systems, reflecting differences in the magnitude of sources and sinks with mesocosm dimension (Berg et al. 1999). Net production and consumption of nitrate were inversely correlated with the surface area:volume ratio ($r^2 = 0.89$, $p < 0.01$; Fig 17). Net nitrate production was highest in dimension D (the darkest tank) followed by dimensions of A, C, and E, while in the small B tanks, nitrate was consumed rather than produced. Differences in light availability played a significant role in determining the fate of nitrate. Tanks with the highest production of nitrate were the deep D tanks which did not receive light at the sediment surface. The smallest range in net nitrate fluxes was observed in the mesocosms with constant

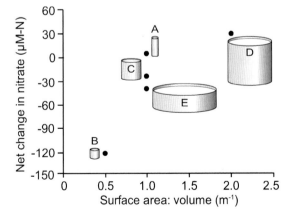

Figure 17: *In MEERC mesocosm experiments, net change in nitrate concentration was inversly correlated with the surface area:volume ratio of the tanks. Positive changes indicate production while negative changes indicate nitrate uptake.*

depth (A, C, E). Nitrate uptake is light dependent, and positive correlations were found between net nitrate uptake and the light at mid-depth of the tanks. Consequently, highest rates of net nitrate uptake were seen in the B tanks.

The attenuation of light is but one factor that impacts the magnitude and direction of nitrogen fluxes. A large difference in the diel water column temperature range was also found among the different mesocosm dimensions, and more than 80% of this difference could be explained by differences in total light availability and the heat emitted from the light source. Furthermore, wall periphyton chlorophyll in the B tanks was found to be up to 16 fold higher than that of the water column chlorophyll in some experiments. Wall periphyton chlorophyll in dimensions C, E, and A, on the other hand, was found to be between three and seven fold that of the water column chlorophyll. These factors, together with feedback effects from changes in trophic interactions, may diverge in different mesocosm systems as surface area:volume is altered, and this in turn leads to variable pathways of nitrogen cycling.

Mesocosms are also excellent tools for understanding phytoplankton preferences for nitrogen form. A fundamental concept in plankton ecology is that of new and regenerated production (Dugdale and Goering 1967). New production is based on nutrients new to a system, such as through upwelling in coastal environments, while regenerated production is that based on nutrient regenerated *in situ*. Nitrate is thus generally considered a "new" form of nitrogen, while ammonium and urea are generally considered to be "regenerated", although it is understood that not all nitrate is new, nor is all ammonium regenerated.

These distinctions become particularly blurred in eutrophic environments where ammonium and urea can be anthropogenic sources as well. Mesocosms provide a tool to understand the fate of these different nitrogen forms, through direct nutrient additions to the tanks, and through the assessment of the regeneration of nutrients that occur through bacterial remineralization and zooplankton release. One of the important relationships that can be directly tested in mesocosms is whether, and to what degree, new nutrients support a different plankton community than regenerated nutrients. It

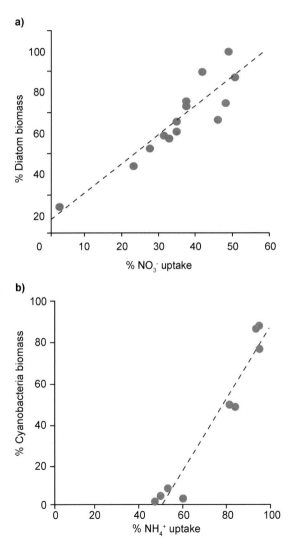

Figure 18: *Phytoplankton community composition, as (a) percent diatoms (b) or percent cyanobacteria was related to the percent uptake of different forms of nitrogen.*

has been hypothesized that shifts in nitrogen form from nitrate to ammonium lead to community shifts away from plankton communities dominated by diatoms to those dominated by flagellates, cyanobacteria, and bacteria, in turn, causing a shift in composition of higher food webs.

Using the MEERC tanks, the relationship between the relative availability of different nitrogen forms and phytoplankton composition was examined in one 4-week experiment. The only nutrients that were supplied were those of natural river water, which was exchanged at the rate of 10% per day. Each of the 5 MEERC tank shapes was used in triplicate for this study. Nitrogen uptake rates were determined using stable isotopic labeling techniques (Glibert and Capone 1993), and phytoplankton composition was assessed using both pigment signatures and microscopic examinations.

The source river water was nitrate-rich, and large diatoms initially dominated the assemblage. Over the time course of the experiment, there was a transition to phytoplankton of smaller cell sizes. Diatoms continued to dominate in the tanks of the constant depth series (dimensions A, C, E). In contrast, in the tanks at both extremes of the constant surface:volume ratio (dimensions B, D), the relative proportion of cyanobacteria increased. The phytoplankton assemblage of the B tanks, by the end of the experiment, was nearly 80% cyanobacteria, while that of dimension D was composed of nearly equal proportions of cyanobacteria and other flagellates.

The rates of uptake of different nitrogen forms were good predictors of the phytoplankton composition in these mesocosms. For all the mesocosms combined, there was a good agreement between the percentage of nitrate uptake and the percentage of diatoms in the phytoplankton community (Fig. 18a). Conversely, where cyanobacteria dominated the assemblages, the percentage of ammonium uptake varied in proportion to the percent cyanobacteria (Fig. 18b).

These relationships have recently been observed in several natural systems. In Moreton Bay, Australia, urea uptake, another measure

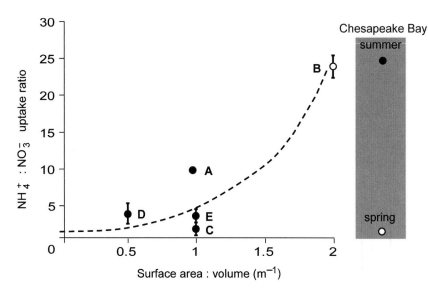

Figure 19: *In MEERC mesocosms, the ratio of uptake of ammonium:nitrate varied with the surface area:volume ratios of the tanks. The ratio of ammonium:nitrate uptake in the large mesocosms (C, D, and E) resembled spring conditions in the Chesapeake Bay; in mesocosm B, this ratio resembled summer conditions in Chesapeake Bay.*

of regenerated production, was also correlated with dinoflagellate abundance (Glibert and Heil 2005), while in Florida Bay, cyanobacterial abundance showed a positive relationship with urea uptake and a negative relationship to nitrate uptake (Glibert et al. 2004). In that system, also, diatom abundance showed a positive relationship to nitrate uptake and a negative relationship to urea uptake.

In eutrophic estuarine and coastal systems, such as the Chesapeake Bay, there is a predictable change in the phytoplankton community size spectrum from larger net plankton (mostly diatoms) in the spring, to small flagellated forms in the summer (Malone et al. 1996). A considerable amount of research on cellular rates of nutrient uptake and community composition have correlated centric diatoms with large and/or frequent additions of nitrate, and microflagellates with regenerated nitrogen additions. Patterns in nitrogen utilization in the larger mesocosms were thus comparable to those in the Chesapeake Bay during spring, whereas patterns of nitrogen utilization in mesocosm dimension B were more comparable to those observed in summer (Fig. 19). Thus, the mesocosms provided a system in which

the dependence on nitrate versus ammonium varied with surface:volume ratio, in contrast to the Chesapeake Bay in which, as succession occurs, the dependence on nitrate versus ammonium varies with season.

Mesocosms can also serve as tools for characterizing the fate of nitrogen in eutrophic systems. Oviatt encapsulated the power of mesocosm experiments by stating, "for experiments in enclosures, all phases and compartments can be measured, such as rates of breakdown, disappearance from the water column, mixing into sediments, and fluxing back from the sediment, perhaps in another form. These data can be used to budget substances accurately under nearly natural conditions allowing the budgets to be extrapolated to the field," (1994).

The fate of nitrogen can be traced in mesocosms in several ways. At the simplest level, an enrichment can be made and the change in concentrations of various dissolved and particulate pools can be measured. A more complex method, but much more powerful, is to enrich with an addition of nitrogen labeled with a stable isotope tracer. In this manner, the pathways of incorporation can be determined even when there is no change in algal biomass–as would be the case when nitrogen is taken up by phytoplankton, but the chlorophyll is also consumed by grazing.

By enriching mesocosms of varying scale with a single pulse of nitrogen labeled with the stable isotope of nitrogen, ^{15}N, the fate and pathways of ammonium were determined in MEERC experiments. These experiments also mimicked an additional aspect of a natural ecosystem:water exchange. On a daily basis, 10% of the water was removed and replaced with riverine water that was not isotopically enriched. Thus, the isotopic enrichment of the ammonium would be expected to decrease even in the absence of any biological or chemical transformation. Comparisons were drawn between the theoretical isotope dilution rate and the observed isotope dilution rate. Any "dilution" in isotopic labeling beyond that predicted by the water exchange reflects biological production of nitrogen (i.e. regeneration).

The "whole tank" labeling approach yielded many insights that would not have been gleaned by monitoring concentration changes alone. Concentration changes showed that ammonium was rapidly consumed. Yet, in all tanks of all size and shape combinations, the dissolved ammonium became depleted in ^{15}N much more rapidly than the rate that would be calculated based on a 10% water exchange per day. Thus, ammonium was being produced by pathways occurring in the tank. In different tanks, the rates of ammonium depletion and production differed. Consistent with rapid depletion of this substrate, in tanks A and B, the smallest volumes, the particulate fraction became rapidly isotopically enriched. It took nearly 24 hours for the particulate nitrogen in the deep tank D to become enriched at the same level. In fact, the tanks with the largest volumes, D and E, continued to show increases in isotopic labeling of the particulate material for up to 48 hours (Fig. 20).

These differences suggest nitrogen was produced differently in tanks of different sizes and that different microbial communities developed. This was further demonstrated by

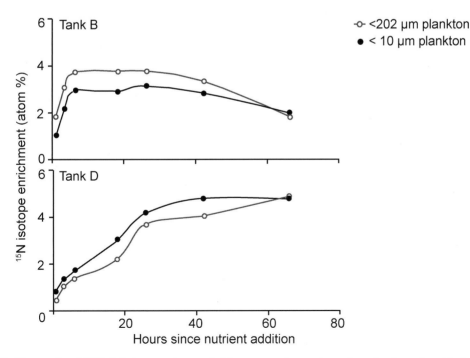

Figure 20: *The uptake of ^{15}N labeled ammonium varied between tanks, as shown here for the tanks with the shallowest (tank B) and deepest (tank D) depths. A more rapid incorporation of ^{15}N ammonium was observed in tank B than tank D, which was dominated by cells that were larger in size than those in tank D.*

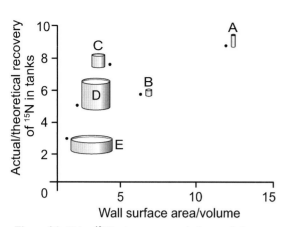

Figure 21: *Using ^{15}N isotopes, a mass balance of nitrogen was tracked in the tanks and compared to what would be expected based on 10% water exchange per day. In all cases, the total amount of ^{15}N labeled nitrogen increased relative to the theoretical calculations showing that these systems were nutrient traps.*

the fact that in the small tanks, the isotopic enrichment in the large plankton fraction (<202 µm) exceeded that in the small plankton fraction (<10 µm), whereas in the large tanks, the isotopic enrichment in the small plankton fraction exceeded that in the large plankton fraction (Fig. 20). A combination of light and wall effects as well as different grazer populations contributed to these differences.

In every experiment in which an isotopically labeled enrichment was made the theoretical calculation of residual nitrogen based on a 10% dilution per day greatly underestimated the actual nitrogen in the mesocosms. In fact, the actual nitrogen in the tanks—as the sum of all fractions of dissolved and particulate nitrogen—exceeded the calculated mass balances by as much as 10 fold, depending on the geometry of the tank (Fig. 21). Thus, in spite of the daily exchange, the mesocosms served as nutrient traps. These mesocosms became model systems for eutrophic estuaries. Nutrients became increasingly trapped in sediments and on the walls and less was bound in water column phytoplankton that could be flushed with the daily exchange. Analogously, in natural estuaries, nutrients become locked in sediments and may continue to serve as a

source even after new nutrients to the water column are reduced.

Management implications

Coastal and estuarine systems around the world are increasingly plagued by outbreaks of harmful algal blooms (HABs, Fig. 22). HABs are those proliferations of algae that either produce a toxin or harmful metabolite that kill or intoxicate fish or shellfish, or that develop to such high biomass that depletion of oxygen occurs as blooms decay, or destruction of habitat occurs from light shading or benthic habitat disruption. Eutrophication is now considered to be one of several factors contributing to global expansion of this phenomenon (Hallagraeff 1993; Anderson et al. 2002; Smayda 1997; Glibert et al. 2005).

Mesocosms are an ideal tool for exploring several specific hypotheses with regard to eutrophication and HABs. As is the case in the development of any particular species, the development of HABs depends on many biological and chemical factors, from nutrients to grazers. The nutrient ratio hypothesis (Smayda 1997; Tilman 1977) states that nutrient loading that leads to enrichments of nitrogen or phosphorus relative to silicate will lead to proportional shifts away from diatom-dominated community, and potentially toward an assemblage more likely to be dominated by HABs. Mesocosms allow for manipulation of enrichment ratios. However, as shown above, mesocosm geometry should also be considered in interpretation of nutrient pathways. Adding to the complexity of the nutrient ratio idea, it has also been recognized that many HAB species can acquire their nutrients in organic form (Romdhane et al. 1998; Berg et al. 1997; Berman and Bronk 2003; Glibert and Legrand 2006), and many are excellent competitors with bacteria for organic substrates. While diatoms are generally nitrate opportunists, autotrophic and heterotophic dinoflagellates may remain competitive under a wider range of nutrient conditions. In MEERC experiments, the phytoplankton community varied with nitrogen form. With increasing proportion of ammonium or urea the community shifted away from diatoms to flagellates or cyanobacteria. With forms of anthropogenic nitrogen shifting globally to more and more organic forms (Glibert et al. 2006), the relationship between nutrient composition, not just total quantity, will be important to further understand if researchers hope to better predict, and ultimately control, HABs (Fig. 22). Many aspects of nutrient dynamics and HABs can be studied in mesocosms.

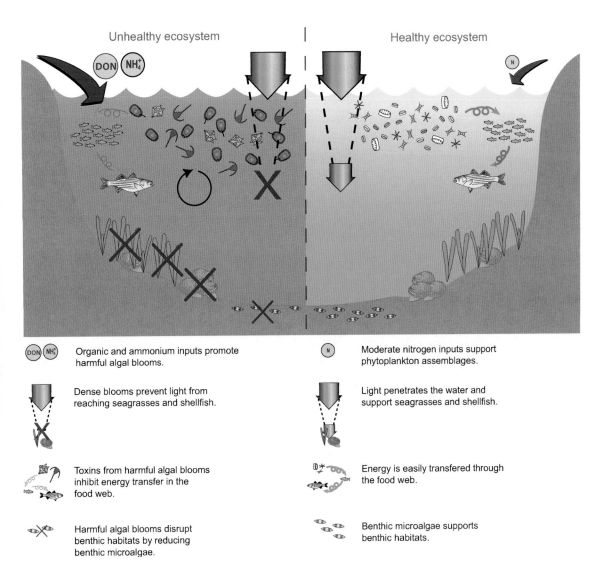

Figure 22: *In a "Healthy" ecosystem, light penetration is high and supports a diverse phytoplankton and benthic community. In an "Unhealthy" ecosystem, light penetration is reduced, there may be greater dependence on "reduced" or "organic" forms of nitrogen, and phytoplankton community composition may shift to flagellates, cyanobacteria, or harmful algal blooms.*

Biofiltration, water quality, and sediment processes
E.T. Porter, J.C. Cornwell, L.P. Sanford, and R.I.E. Newell

Historical oyster populations in places such as Chesapeake Bay used to be able to filter algae and other particles from the water, thus regulating water quality (Fig. 23). Excess nutrient inputs and declines of oyster biofiltration have led to more turbid water, reducing the growth of benthic microalgae and affecting the exchange of nutrients between sediments and the overlying water. Oyster densities and bottom shear velocities were manipulated in mesocosm experiments to investigate their effects on benthic microalgae, nutrient regeneration from sediments, and overall water quality in the ecosystem. The mesocosm results demonstrate that biofiltration removes particles from the water

column, allowing light to reach the bottom, and stimulating benthic microalgal growth, which in turn stabilizes sediments and decreases nutrient release into the water column. Realistic shear velocities have the potential to erode these benthic microalgae; however, exerting an additional physical control on benthic biogeochemical exchanges. Even considering this physical limitation, the results indicate that management efforts to increase biofiltration (e.g., oyster restoration) will have multiple, synergistic positive ecosystem outcomes.

Problem description
Over the past half-century, water quality and transparency have declined in many eutrophic

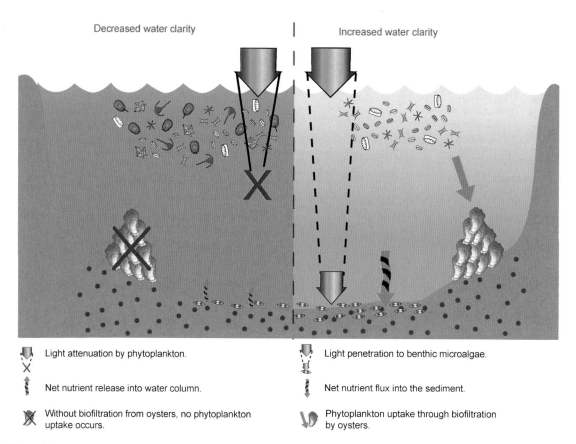

Figure 23: *Healthy oyster populations can improve water quality and clarity.*

estuaries. One such estuary, Chesapeake Bay, has also experienced a concurrent decline in the abundance of the eastern oyster, *Crassostrea virginica*, which formerly supported a productive fishery. Management authorities in Chesapeake Bay and elsewhere have recommended that water quality conditions and fishery harvest could both be improved by restoring biomass of the eastern oyster to a modest fraction of historical levels (EPA 2000). In shallow-water environments, benthic and pelagic processes are closely coupled and water flow can regulate the supply of seston to bivalves. In addition, such flow may regulate water clarity and benthic-pelagic nutrient fluxes through mixing and resuspension of bottom sediments. Until recently, the complex interactions between oysters, nutrients, water clarity, and physical circulation were poorly understood.

MEERC researchers designed a series of studies using experimental mesocosms (Fig. 24) to quantify how the combined effect of oysters and increased bottom shear velocity directly or indirectly affect ecosystem processes and shift ecosystem function between the water column and sediments (Porter et al. 2004a, 2004b). Oysters and bottom shear velocity were used to examine effects on the following factors:

- Phytoplankton abundance in mesocosm experiments;
- Nutrient transformations and nutrient regeneration from the sediments;
- Overall water quality in whole-ecosystem experiments.

To address these questions, the interacting effects of juvenile oysters and bottom shear velocity on phytoplankton biomass and on nutrient regeneration in a series of three (spring, summer, fall) 4-week long mesocosm experiments were studied under different levels of bottom shear (Porter et al. 2004a, 2004b). The mesocosms were 1-m deep, had a 1 m^2 sediment surface area, contained 1000 L of water, and received the same water-column mixing designed to simulate conditions in nature (turbulence intensity 1 cm s^{-1}) (Porter et al. 2004a, 2004b). A parallel set of smaller (0.1 m^3) experimental systems was used for comparative studies (Porter

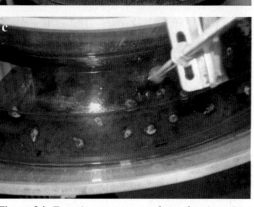

Figure 24: *Experiments were performed to investigate how oyster populations in interaction with low and moderate bottom shear affect water quality. (a) Healthy oysters can improve water clarity through biofiltration. (b) A large linked mesocosm with an annular flume was designed and an experiment conducted to study the interaction of water flow and oysters on water quality. (c) Oysters were placed in the annular flume and water column mixing and bottom shear scaled in comparative systems.*

et al. 2004a). Experimental systems included a multicomponent mesocosm with moderate bottom shear velocity (0.6 cm s^{-1}) and two standard cylindrical tanks with an unrealistically

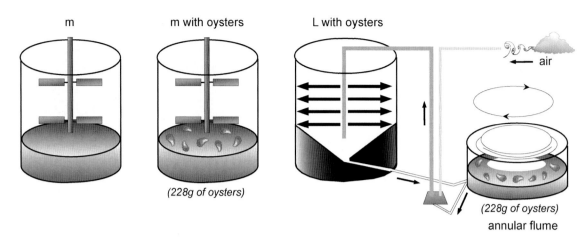

Figure 25: *Experimental ecosystems with and without oysters and low and moderate bottom shear. All mesocosms were 1 m deep and contained 1000 L of water (the linked mesocosms together contained 1000 L). Two mesocosms and the annular flume had 1 m² of sediment surface area. Description of treatments: m: mesocosm with low bottom shear and without oysters; m and oysters: mesocosm with low bottom shear and with oysters; L with oysters: linked mesocosm and annular flume with moderate bottom shear and with oysters. Additional experiments were conducted using a similar system of mesocosms with and without realistic benthic boundary layer shear, but with water column volumes of 100 L, 2004b).*

low bottom shear velocity (0.1 cm s⁻¹, Fig. 24 and Fig. 25). Experiments were run with and without juvenile oysters, using oyster densities similar to oyster abundances in historic times (nineteenth century) in Chesapeake Bay (Newell 1988).

Research findings

It was found that direct and indirect interactions between oysters and moderate bottom shear velocity affected phytoplankton biomass, light availability in the water column and at the sediment surface, microphytobenthos biomass, and nutrient regeneration from the sediments to the water column. Oyster feeding significantly decreased phytoplankton biomass. The isolated tank without oysters repeatedly developed a phytoplankton bloom while the mesocosms with oysters did not. The oyster-mediated decrease in phytoplankton biomass also consistently led to enhanced light penetration through the water column to the sediments (Porter et al. 2004a, 2004b)

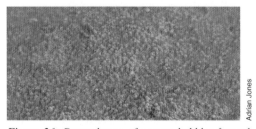

Figure 26: *Dense layers of oxygen bubbles formed from a benthic algal mat (above). In mesocosm experiments with moderate bottom shear these bubbles increase erosion of the microphytobenthos mat. Sediment chlorophyll a concentrations as indicator of microphytobenthos biomass at the end of experiments 2 and 3 (right panel). Different letters indicate statistically significant differences between treatments. Treatments m+oysters and L+ oysters had the same biofiltration (i.e. increased light levels at the sediment water interface); however, a shear velocity of 0.6 cm s⁻¹ eroded microphytobenthos (Porter et al. 2004a, 2004b).*

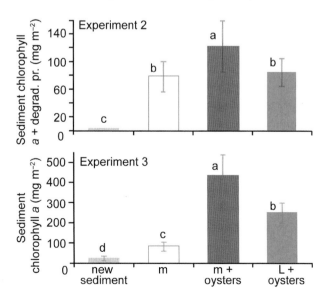

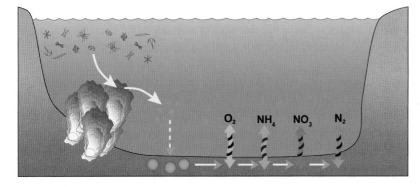

Figure 27: *After consuming phytoplankton, oysters release biodeposits. These biodeposits begin the process of nutrient cycling in sediments. Benthic microalgae take up nutrients regenerated from the biodeposits and coupled nitrification–denitrification takes place. This cycling was mimicked in experiments in benthic chambers (see Fig. 28).*

Light availability at the bottom enhanced benthic microalgal biomass (Fig. 26), thereby reinforcing feedback effects to retard sediment-water fluxes of nutrients. There was a significant increase in the daily sediment uptake of dissolved inorganic nitrogen with increasing sediment chlorophyll *a* abundance. Thus, benthic microalgae significantly reduced the overall amount of regenerated nitrogen that was returned to the water column. The daily nitrogen release to the water column was lowest in the system with oysters and low bottom shear (m with oysters, Fig. 25) which generated the highest microphytobenthos biomass.

In these experiments (Porter et al. 2000a, 2004b; Porter 1999), well-developed benthic microalgal communities often formed cohesive microphytobenthos mats (Fig. 26) and microphytobenthos has been known to stabilize sediments (Madsen et al. 1993).

However, toward the end of these 4-week long experiments, bubble formation within these mature benthic algal mats tended to increase their buoyancy and benthic friction (roughness), thereby making them more susceptible to erosion by bottom shear. Consequently, experiments in systems with moderate bottom shear velocity ($0.6\,\mathrm{cm\,s^{-1}}$) exhibited substantial erosion of benthic microalgae (Fig. 26), which resulted in higher benthic nutrient recycling, despite increased light availability due to oyster feeding. The sediments in the experiments described above were fine-grain (mud). These studies emphasize the importance of considering both benthic feeding and bottom shear on water quality.

Subsequent independent benthic chamber experiments demonstrated that the addition of particulate organic matter to simulate oyster biodeposits (Fig. 27) also significantly altered sediment nutrient cycling processes (Newell et al. 2002). Experimental biodeposition of organic matter simulating oyster processes led to small increases in ammonium recycling from sediments to overlying water (Fig. 28). However, rates of denitrification were greatly stimulated by biodeposition, resulting in a large net removal of available nitrogen from the water column (Fig. 28). In estuaries such as Chesapeake Bay, this effect might ultimately lead to further reductions in phytoplankton biomass because of nitrogen limitation of cell growth.

Management implications

Results of these studies indicate that filter-feeding bivalves, in conjunction with water flow and bottom shear, can affect pelagic-benthic processes through a range of complex interactions. Bivalve filtration causes decreased phytoplankton biomass and increased water clarity. Clearer water promotes growth of benthic microalgae that tend to cap nutrient recycling fluxes from sediments to overlying water. Reduced benthic nutrient recycling further retards growth of phytoplankton. Biodeposition of organic matter from oyster filtration increases nutrient delivery to the sediments but also enhances bacterial removal of fixed nitrogen through denitrification. These processes further reduce nutrient availability for sustaining phytoplankton growth. Benthic microalgal communities excrete mucus that tends to bind sediment particles together, making them less susceptible to resuspension and therefore helping to maintain clearer water.

Under some conditions benthic microalgae form mats that become buoyant when bubbles of photosynthetically-produced oxygen become trapped in the algal matrix. The erodablity of bottom sediments will thus depend both on bottom shear velocities and on the nature of the benthic microalgal community. Bottom shear velocity in Chesapeake Bay is higher (1.0–$1.4\,\mathrm{cm\,s^{-1}}$) than the moderate bottom shear ($0.6\,\mathrm{cm\,s^{-1}}$) used in this mesocosm experiment, which likely causes additional erosion of benthic microalgae.

Experiments and models designed to aid in prediction of the effects of bivalve suspensions-feeders and water flow on ecosystems must include realistic physical conditions. Data used in models must come from experiments that consider direct and indirect effects of interactions between biological and physical components of the ecosystem. Controlled mesocosm experiments that vary physical variables such as bottom shear and sediment type and biological variables such as bivalve species and density are needed to further resolve the complex interactive effects of bivalves and water flow on benthic-pelagic coupling and on overall water quality. Specifically, MEERC researchers suggest the need for (1) designing a new generation of mesocosms with realistic water-column turbulence levels and high bottom shear stress, (2) conducting comparative ecosystem studies with tidal or episodic sediment resuspension and additional benthic species, and (3) considering effects of sediment type and age of benthic microalgal community on overall benthic-pelagic dynamics.

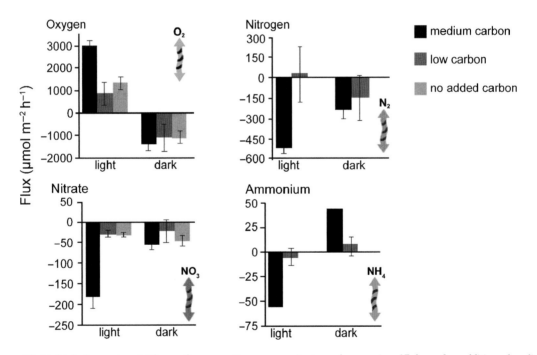

Figure 28: *Light:dark experiment. Fluxes of oxygen, nitrogen gas, nitrate, and ammonium 17 days after addition of medium, low, and zero amounts of particulate organic matter, "Carbon," were added to benthic chambers. The bars indicate the means ± standard deviations). As indicated on the x-axis, sediment fluxes on the cores were determined in the light and in the dark. Positive values indicate a flux out of the sediment to the overlying water; negative values indicate flux into the sediment. Control cores only containing water to check for water column processes exhibited minor fluxes that do not show at the scale of these figures.*

Food webs, trophic efficiency, and nutrient enrichment
W. M. Kemp and M.T. Brooks

Nutrient enrichment to coastal oceans affects not only the primary producers (e.g,. algae) that absorb the nutrients, but also the food web that connects bacteria, herbivores, and carnivores, including diverse fish species. The efficiency of energy transfer from lower trophic levels (e.g., phytoplankton) to higher trophic levels (e.g., zooplankton and fish) was investigated with field data, mesocosm research, and modeling approaches. Results from these experiments and models indicate a reduction in trophic efficiency (ratio of fishery catch per unit primary production) with increasing nutrient loading. Thus, nonlinear relationships between nutrient loading and fish production occur, with initial increases in nutrient loads stimulating fish production, but further increases leading to declines. Proposed nutrient reductions in many coastal waters are likely to have beneficial outcomes of improved water quality and trophic efficiencies without detrimental impacts on fisheries production.

Problem description

Coastal eutrophication is a growing problem in estuaries, lagoons, and embayments throughout the world. Nutrient enrichment tends to stimulate growth and abundance and alter the composition of algal communities, often leading to major changes in ecosystem structure (Fig. 29). In shallower waters, eutrophic conditions can cause loss of fish and invertebrate habitats in seagrass beds because accumulation of algae in the water column and on leaf surfaces tends to shade, and thereby inhibit growth of, seagrass plants. Animal habitats in deeper waters can also be degraded

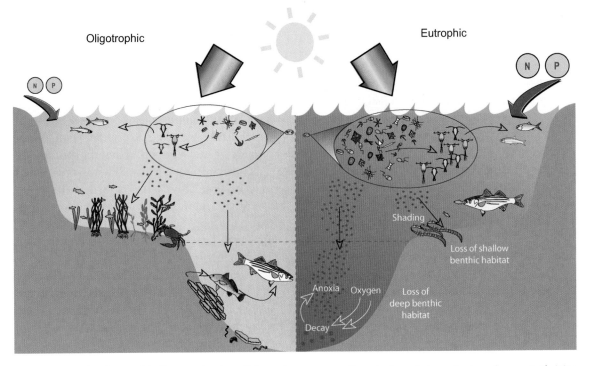

Figure 29: *Under oligotrophic (low nutrient) conditions, clearer water allows bottom photosynthesis and supports thriving benthic habitats for fish and invertebrates. Under eutrophic conditions, high growth of algae in the water tends to support greater biomass of plankton, but reduces water clarity and stimulates depletion of oxygen from bottom waters, thereby degrading bottom habitats and associated fish.*

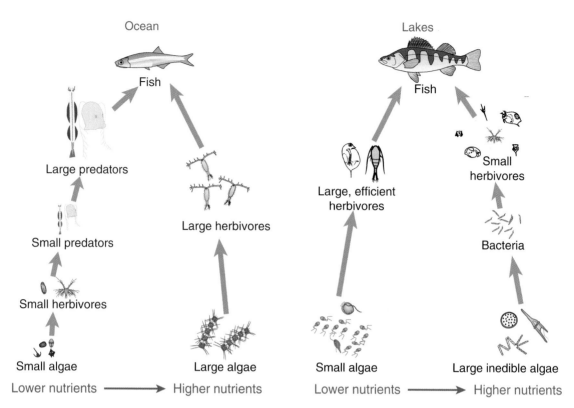

Figure 30: *Food chains are altered with nutrient enrichment in oceanic and lake waters. With natural enrichment of nutrients from upwelling of deep waters, shorter oceanic food chains supported by large planktonic algae tend to be more efficient at producing fish. Nutrient enrichment of lake waters, however, removes efficient herbivores, causing food chains to be longer and less efficient at producing fish.*

by eutrophication by which nutrients stimulate growth of phytoplankton that sinks to and decays in bottom waters, causing rapid consumption of dissolved oxygen and, in some instances, leading to complete oxygen depletion (anoxia).

Nutrient enrichment tends to change the nature of major aquatic food chains that link phytoplankton production to fish growth. In the inherently oligotrophic open ocean, coastal upwelling of nutrients stimulates growth of larger algae that are consumed by larger zooplankton, which are, in turn, eaten directly by fish (Fig. 30). Under normal low nutrient conditions, ocean plankton food chains are long, starting with small algal cells and often passing through gelatinous predators before reaching fish consumers (Landry 1977). Plankton food chains in lakes may be structured differently, with low-nutrient lakes often supporting large zooplankton (cladocerans) that efficiently filter small algal cells to create short food chains linking phytoplankton to fish. Nutrient

enrichment of lakes, however, tends to produce longer planktonic food chains, which are less efficient. In this case, large algal cells at the base of eutrophic lake food-webs are toxic or inedible, and their excretion of dissolved organic matter supports bacterial production which is consumed by small herbivores, which may be eaten by fish (Hillbricht-Illkowska 1997). Depending on environmental conditions, estuarine and coastal plankton food-chains appear to have characteristics of both oceanic and lake ecosystems.

Coastal systems are thought to pass through three stages (Fig. 31) as they become progressively more eutrophic (Caddy 1993). In the first stage, nutrient enrichment tends to increase food supply thereby supporting higher rates of fish production (Luo and Brandt 1993). In the second stage, fish and other consumers in the upper levels of food chains are no longer limited by food supply and fish production is controlled by other factors such as climate and fishing mortality. In the third stage, nutrient enrichment leads to degradation of crucial habitats that are needed by early life-stages or by other physiological requirements of fish (particularly, bottom-dwelling fish, de Leiva Moreno et al. 2000). Consequently, further nutrient additions lead to decreases in fish production. This model suggests that the response of fish populations to changes in nutrient loading will depend on which stage characterizes their particular coastal ecosystem. Although many eutrophic coastal systems also suffer from declining fisheries production, the contribution of nutrient-related shifts in trophic interactions to these trends is unclear.

In many coastal systems worldwide, large-scale efforts are underway to improve water quality conditions by reducing nutrient loading. In most instances, there has been surprisingly little attention given to the question of how lowering of nutrient inputs will affect fish production and harvest in these systems. For systems that

are in Stage 1 of Caddy's model (Caddy 1993), reduced nutrient loading may cause a decrease in fish production. For Stage 2 systems, modest increases or decreases in nutrient loading will have little effect on fish. However, for Stage 3 systems, decreases in nutrient loading may actually increase fish production (Fig. 31). Because of the vast social and economical costs associated with reducing nutrient loading, it is crucial that the scientific community addresses this difficult question. Trophic network models and ecosystem simulation models were developed in conjunction with MEERC studies. This modeling was motivated initially by interest in better understanding production, nutrient cycling, and fish bioenergetics in relation to mesocosm experiments. Questions posed by Chesapeake Bay Program managers regarding how fish production might respond to reductions in nutrient loading led to the application of the models to broader issues.

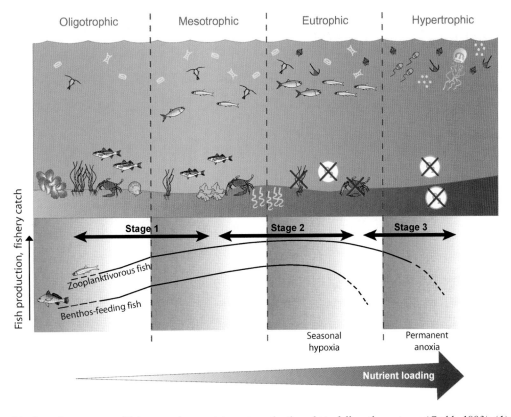

Figure 31: Growth responses of fish to nutrient enrichment can be thought to follow three stages (Caddy 1993): (1) initial nutrient increases stimulate fish growth; (2) subsequent nutrient increases do not affect fish growth; (3) further nutrient increases cause fish to decline, as benthic habitats are lost.

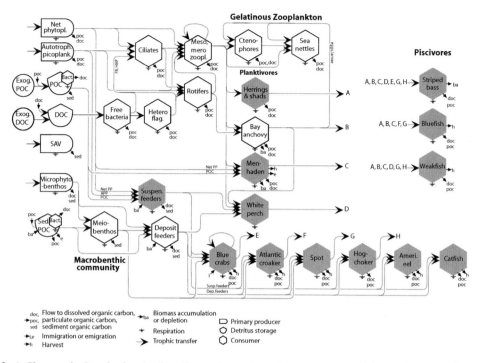

Figure 32: *A Chesapeake Bay food-web where flows of organic carbon are connected from primary producers through intermediate trophic levels to support fish, including 14 consumer populations that constitute 85–95% of the annual fisheries harvest (Hagy 2002). This model allows researchers to quantify the amount of primary production supporting individual fish populations, the number of trophic links from primary producers to each fish population, and the extent to which the population's production depends on benthic habitats.*

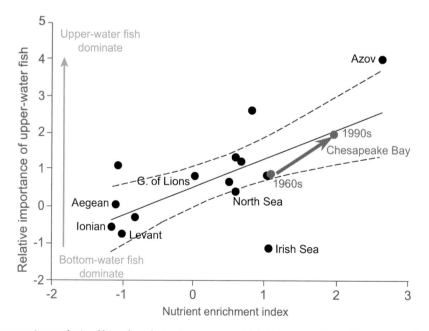

Figure 33: *Comparative analysis of how the relative importance of fish living near the sediment–water interface changes with nutrient enrichment for a diversity of coastal ecosystems and for Chesapeake Bay in the 1960s versus the 1990s (modified from de Leiva Moreno et al. 2000).*

Research findings

Models developed to analyze experimental ecosystem dynamics have been used to quantify food-web interactions and analyze trophic responses to nutrient enrichment (Hagy 2000). Trophic network models have been used for calculating: (1) the average trophic level (number of feeding links from plants to consumers) for each consumer population, (2) the fraction of each animal's diet that passes through plankton-only pathways (i.e., no benthic organisms are included), and (3) the trophic efficiency by which primary plant production is converted to production of a given consumer population. Trophic networks were developed to portray feeding relationships in Chesapeake Bay, including 14 specific consumer populations that account for 85–95% of the total fisheries biomass harvest in the estuary (Fig. 32). Based on these model calculations, percentages of food consumption passing only through plankton pathways were computed for each harvested population and the total harvest was partitioned into that fraction passing through planktonic versus benthic habitats. The computed ratio of plankton-to-benthic fisheries production increased significantly over time as a nutrient enrichment index (log of phytoplankton chlorophyll concentration) from the 1960s to the

1990s. This trend, which is consistent with that reported previously from comparative analysis of different coastal ecosystems, indicates that eutrophication tends to shift trophic pathways away from benthic habitats in favor of habitats in the upper water column (Fig. 33). Similar analyses suggest that the average trophic level from which Bay fisheries are harvested has been decreasing during this same time interval (Pauly et al. 1998; Kemp unpublished). Analysis of historical data on phytoplankton and fisheries indicate that the average trophic efficiency (ratio of fish harvest to phytoplankton production) for many harvested species has been declining during the last five decades. This trophic efficiency also shows a significant inverse relationship with nutrient loading (nitrate, which is a proxy for total nitrogen) for Atlantic menhaden, which are primarily herbivores; for blue crabs, which have an average trophic level of first carnivores; and for striped bass, which feed between first and second carnivore trophic levels (Fig. 34). In general, it appears that the slopes of these relationships become steeper and the variance increases as the organism's mean trophic level increases.

In addition to trophic network models, simulation models were also used to analyze dynamics of experimental pelagic ecosystems

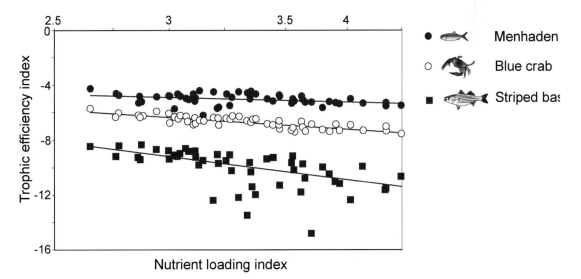

Figure 34: *Response of trophic efficiency (ratio of fishery catch per unit primary production) to increases in nutrient loading for three valuable fisheries populations in Chesapeake Bay. Note the steeper but more variable decline for striped bass feeding at a higher trophic level compared to herbivorous menhaden.*

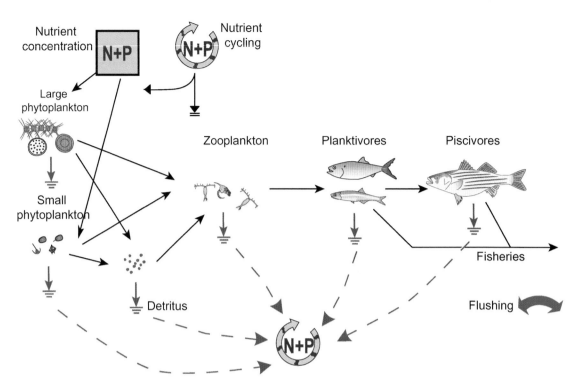

Figure 35: *Simplified food web diagram depicting the variables and processes included in a simulation model to investigate coastal ecosystem response to nutrient enrichment.*

and to investigate responses to nutrient enrichment (Kemp et al. 2001; Hagy 2002; Brooks 2004; Bartleson et al. 2005). These relatively simple models (Fig. 35) included variables for large and small phytoplankton cells, dissolved and particulate detrital organic matter, bacteria, protozoa, and copepods, as well as two different fish groups—those feeding on algae and copepods (e.g., menhaden and bay anchovy), and those feeding on fish (e.g., striped bass, bluefish). All simulation experiments, using a variety of food web configurations tested over a range of nutrient loading and fish harvest conditions, exhibited similar patterns. As long as simulations included significant mortality for consumers at the upper trophic levels, biomass of algae tended to increase over a wide range of nutrient loading rates while zooplankton and fish abundance saturated (i.e., leveled off) at moderately low loading. In this case, top-down control associated with predation tended to limit the maximum biomass at some lower or intermediate nutrient loading rate for all consumer groups while still allowing phytoplankton to continue growing

(Fig. 36a). As a consequence, the trophic efficiencies (animal production/food production) for fish and copepods tended to decline with increasing nutrient input (Fig. 36b). Compared to the conceptual model described earlier (Caddy 1993), it appears that these simulations capture the dynamics of Stage 1 where biomass and production of fish and other consumers increases with rising nutrient input rates at low nutrient levels and Stage 2 where consumers remain relatively constant as nutrient inputs increase beyond a relatively low to moderate level. Because this simulation model does not include benthic habitats, which are susceptible to degradation trajectories that are associated with nutrient enrichment, it does not generate Stage 3 conditions.

Management implications

These modeling studies demonstrate that under certain conditions increasing nutrient inputs can drive coastal ecosystems into Stage 3 where habitat degradation causes fish abundance and production to decline with further nutrient enrichment. Use of network models to analyze fisheries data from Chesapeake Bay and other coastal ecosystems shows that nutrient enrichment can cause fish community structure to shift strongly from dominance by bottom-dwelling species to dominance by species residing in the upper water column.

Simulation experiments also suggest that Stage 2 conditions—where fish production is not limited by food—may characterize many coastal ecosystems today, particularly systems adjacent to cities and rivers that drain agricultural landscapes. Simulation experiments and trophic analysis of historical data both suggest that, while further nutrient enrichment of Stage 2 systems may have little effect on fish abundance, trophic efficiency (fish production per phytoplankton production) tends to decline.

This implies that, in Stage 2, excess primary production that does not go toward supporting fish growth leads to other responses such as increased algal blooming and decreased water clarity. These responses alter environmental conditions but affect fish populations only through indirect interactions.

Comparative analyses of different coastal systems have revealed positive relationships between fish harvest and phytoplankton production (Nixon and Buckley 2004). Such relationships imply that these ecosystems are generally in Stage 1. They further suggest that widely reported ecological changes associated with eutrophication (e.g., hypoxia, seagrass loss, and harmful algal blooms) have little effect on total fish production. These

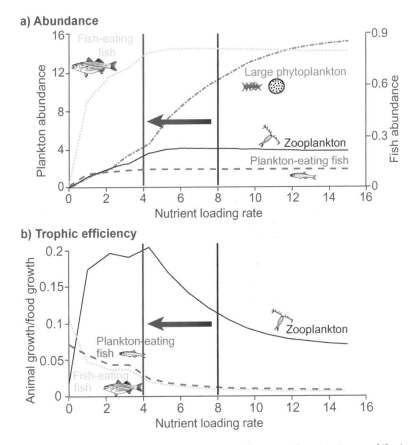

Figure 36: *Model simulations show that algae keep growing over a wide range of nutrient inputs while simulated abundance of zooplankton and fish level off at lower nutrients (upper panel). Trophic efficiency (amount of animal growth supported by a unit of algal production) declines as the animal population levels off (lower panel). Note that a 50% reduction in nutrient inputs (from 8 to 4), would shift the model ecosystem to the left (from red to blue line), causing a dramatic decrease in algae but virtually no change in fish.*

relationships between phytoplankton and fisheries tell us that systems that have higher inherent characteristics can support higher algal production also tend to support more fish growth (Kemp et al. 2005). Such relationships, however, say little about how fish production for any particular coastal system will change with increases or decreases in nutrient loading. In fact, coastal systems with relatively high water residence times and shallow depth tend to have greater capacity for sustaining high rates of primary production; however, they are also more susceptible to habitat-degradation effects of eutrophication.

The simulation model results allowed a consideration of the potential consequences of reducing nutrient loading to coastal systems. In Chesapeake Bay, for example, it has been suggested that a 40% reduction in nutrient (N and P) inputs would be required to achieve the target improvements in water quality (Malone et al. 1991). Although it is not known exactly how to scale the model simulations to actual nutrient loading for Chesapeake Bay, it is estimated that intermediate rates (e.g., 8 µmol L^{-1} d^{-1}) are representative of current conditions

in several Bay tributaries. Model output indicates that nutrient input reduction of 50% would reduce algal biomass by more than 60% while causing no decline in planktivorous (plankton-eating) fish abundance and only 5% reduction in piscivorous (fish-eating) fish. Presumably, the large reduction in phytoplankton biomass would help to reduce hypoxia, improve water clarity, and rehabilitate seagrasses, as well as increasing overall trophic efficiencies, with little or no negative impact on fish.

These model and data analyses indicate that proposed reductions in nutrient loading to estuaries such as Chesapeake Bay will probably improve water quality conditions, help restore populations of some bottom-dwelling fish, and improve overall trophic efficiencies (Fig. 37) while having little effect on total fish abundance and production. However, caution is required because reduced nutrient loading might result in unanticipated non-linear trajectories for ecosystem recovery and shifts in ecosystem structure is required. Such patterns and trends have been seen in other coastal environments undergoing major reductions in nutrient loading (Yamamoto 2003).

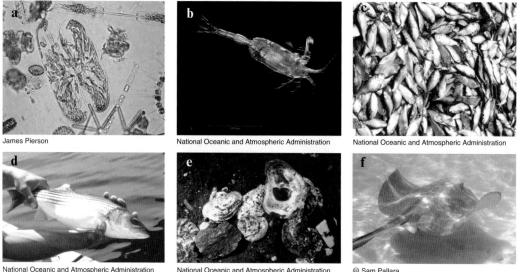

Figure 37: Organisms representing major trophic groups: (a) phytoplankton, (b) copepod, (c) menhaden, (d) striped bass, (e) oysters, and (f) sting ray.

Toxicity and bioaccumulation in benthic organisms
R.P. Mason and E.T. Porter

Toxic chemicals that enter coastal oceans accumulate in the water, sediments, and biota (Fig. 38). Sediments act as a repository for toxins and exchange processes between sediments and the overlying water affect the bioavailability of toxins for organisms. Mesocosm experiments are particularly important for toxicant studies because of the various feedback mechanisms between water, sediment, and biota. Mesocosm results combined with modeling results demonstrated the importance of organic matter binding of toxicants. For example, higher organic matter content of sediments correlated with less methylmercury bioaccumulation in zooplankton. The implication of these results is that food web dynamics and eutrophication status in coastal waters has a larger impact on toxicant dynamics than physical processes such as sediment resuspension. However, sediment resuspension may affect the ecosystem structure. Toxicant bioaccumulation
needs to be monitored closely when restoration efforts result in changes in nutrient loading.

Problem description

The coastal zone has been highly affected by human activities, which has resulted in a large insult of chemicals being introduced into these fragile ecosystems. Estuarine sediments may contain a complex mixture of contaminants because estuaries are often near urban areas and may have received substantial inputs over time from human activities, either by direct discharge or from runoff from the watershed (EPA 2004; Bianchi 2007). The intense physical mixing of the water column and the strong benthic-pelagic coupling that exists in estuaries has important effects on contaminant fate and burial within the sediment. Sediment burial is often the main removal mechanism of pollutants from the system. Consequently, sediments are the long-term sink

 Mercury enters the coastal ecosystem through industrial emissions, and inputs from local sources, as well as through inputs from the watershed.

 Mercury accumulates in wetlands and sediments where it is converted to methylmercury, the form taken up by other organisms.

 Mercury moves through the water cycle: mercury deposits with precipitation and is lost to the atmosphere via gas exchange.

Methylmercury moves through the food chain from plankton to fish to other predators and humans.

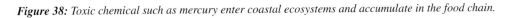
Figure 38: *Toxic chemical such as mercury enter coastal ecosystems and accumulate in the food chain.*

for many chemicals of concern, such as mercury, polycyclic aromatic hydrocarbons (PAHs), and polychlorinated biphenyls (PCBs), because of their toxicity to and bioaccumulation by aquatic organisms, birds, and mammals, including humans (EPA 2004; Bianchi 2007).

Because sediments act as a repository for contaminants in coastal systems, processes that govern the exchange of chemicals between the water, sediment, and biota will have a large effect on the ecosystem (Fig. 38). Sediment burial will eventually remove these contaminants from interacting with the biota in the ecosystem but this is a long-term process. Sediment organic carbon content and sulfide levels play an important role in metal sequestration such that metals are more available for bioaccumulation, are more toxic, and are released to the overlying water more readily in low organic content sediments than in highly organic-rich media (Di Torro et al. 2005; Mason 2002). However, many contaminant concentrations increase with increasing organic content and thus there is a complex interaction between organic matter content and contaminant bioaccumulation from sediments, especially for trace metals. Even though government regulation may reduce contaminant inputs from point and diffuse sources such as watershed sediment loading and agricultural runoff, it is still important to understand the cycling of contaminants within the ecosystem, and particularly within the sediment, because of the potential for release back to the ecosystem, and for their bioaccumulation and toxicity to benthic organisms.

Nutrient loads and system productivity in the water column also influence the concentration and growth dynamics of primary producers and other microbial organisms, and this dictates to a large degree the contaminant concentration at the base of the food chain. Thus, contaminant fate is influenced by factors such as the degree of eutrophication, sediment resuspension, and other ecosystem disturbances. The interactions are often non-linear, and may have secondary effects and feedback interactions (Fig. 39). Thus these complexities cannot be examined in small-scale microcosms or in beakers. In addition, many contaminants are adsorbed to container walls, or by the microbial growth that often forms on the walls of experimental systems during long-term studies. Thus, small microcosms may experience substantial wall effects associated with their use

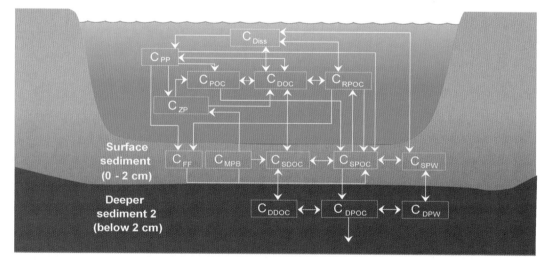

Figure 39: Conceptual model showing the interactions and pathways for a chemical, C, cycling between the sediment and water under the influence of sediment resuspension and the interaction of the chemical with the various phases in the system. The following abbreviations are used: PP phytoplankton, ZP zooplankton, POC particulate organic carbon, DOC dissolved (colloidal) organic carbon, RPOC resuspended particulate matter, Diss dissolved constituents, FF filter feeders, MPB microphytobenthos, S surface sediment, D deep sediment, PW porewater.

Figure 40: The MEERC STORM (Shear, Turbulence Resuspension Mesocosm) system was used to investigate physicochemical processes, sediment–water interactions, and transfer of contaminants between media. The system was more effective at studying benthic-pelagic coupling processes than typical mesocosms because the STORM system mimicked realistically episodic and tidal resuspension with high bottom shear and realistic water column turbulence levels, without over-stirring the water column.

due to their large wall-surface-area-to-volume ratio. Such wall effects are more important in the absence of sediment in the experimental ecosystem. Experiments that include sediment are the most realistic for the examination of the effect of contaminants and toxicants on coastal ecosystems, and on contaminant bioaccumulation and health effects on fish consumers.

Sources of contaminants (from water, sediment, or food) to biota cannot be determined from field studies or simple in-lab exposure experiments because of the importance of feedback mechanisms and the interaction between species (Fig. 39). Recent studies have shown that mesocosms have substantial advantages for the examination of contaminant biogeochemical cycling within complex systems such as estuaries (Orihel et al. 2006; Bromilow et al. 2006). Benthic-pelagic coupling by physical (advection, diffusion, and resuspension), chemical (sediment redox changes), or biological (biota migration across the sediment–water interface) processes can strongly affect the rate of bioaccumulation. System growth rate and productivity influenced by nutrient loadings also influences both

organism feeding rates and bioaccumulation (growth dilution effects). Finally, uptake of contaminants may be slow such that longer-term experiments need to be performed over weeks to months to ascertain clearly the uptake mechanisms and the effects of contaminants and their potential toxicity. Again, small-scale systems cannot maintain their integrity for sufficient time for many of the studies that need be done to examine bioaccumulation and trophic transfer of contaminants in aquatic food chains. Studies (Kim 2004; Kim et al. 2004; Kim et al. 2006) have demonstrated, for example, the importance of longer term experiments for examining estuarine mercury cycling.

Overall, the real world is complex and requires detailed, long-term experiments to adequately examine all the interactions in systems with realistic physical mixing and disturbance and representative food chains. Many important questions cannot be examined through the collection of field data. Therefore, mesocoms provide a system that can be manipulated to examine the complex processes discussed above in long-term controlled experiments. Limitations of size need to be heeded, especially if higher level food chain organisms are to be included. A typical 1 m^3 mesocosm would not be suitable for fish studies except with the smallest fish. However, the typical mesocom size is suitable for the examination of the effect of contaminants on invertebrates, both benthic and pelagic. Benthic-pelagic coupling studies have recently become more realistic by using experimental ecosystems with realistic bottom shear and water column turbulence (Porter et al. 2004b).

Research findings

Mesocosm studies using the MEERC STORM system (Fig. 40) have been used to investigate purely physicochemical processes in relatively simple systems, and to examine sediment–water interactions and the diffusive and advective transfer of contaminants between media (e.g., water and sediment; Schneider 2005; Schneider et al. 2007). In addition, they have been shown to be suitable for examination of the bioaccumulation of contaminants in a relatively complex system with substantial benthic-pelagic coupling and where there are both water-column and sediment-based primary consumers (adapted from Kim et al. 2004, 2006, in review; Kim 2004). More specifically, they have been used to examine the role of physical disturbance, such

as tidal and episodic sediment resuspension, on contaminant transport from sediment to the water column, and on bioaccumulation by biota for mercury (Kim 2004; Kim et al. 2004, 2006; Bergeron 2005).

MEERC STORM systems of $1m^3$ volume were designed for mesocosm experiments that include episodic and tidal resuspension in short-term and longer-term mesocosm experiments. Episodic resuspension due to increased bottom shear can be induced by storms yet its effects on the ecosystem, the nutrient, and the contaminant cycling are difficult to assess in the field. In addition, bottom shear varies over the tidal cycle. Sediment resuspension is induced regularly during the tidal cycle when critical erosional thresholds are surpassed, such as, during highest flood or ebb tides.

Bottom shear was carefully controlled in the STORM systems and set to levels of bottom shear found in the field and that induced sediment resuspension. However, at all times water column turbulence levels were kept at realistic levels and the water column was not overmixed. Bottom shear in standard mesocosms is unrealistically low, with consequences for benthic-pelagic coupling (Porter et al. 2004b). In the STORM systems, using a special mixing apparatus, much higher bottom shear without overmixing the water column can be induced.

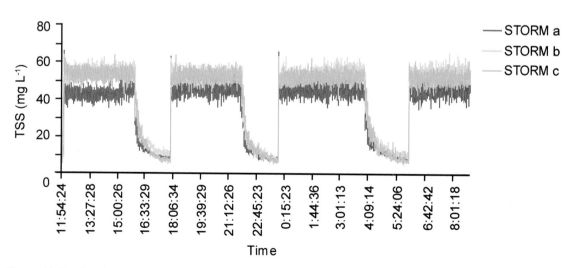

Figure 41: *Tidal cycles in a set of triplicate STORM tanks induce sediment resuspension during three 4-hour mixing_ON phases and induce particle settling during 2-hour mixing_OFF phases. TSS = total suspended solids.*

Bottom shear in the STORM systems can be programmed to vary smoothly over any desired cycling pattern. For logistical reasons related to sampling a large number of ecosystem variables, the system mimicked tidal resuspension as 4-hour "mixing on" periods with high bottom shear during which sediment resuspension occurred, and a 2-hour mixing off phase (Fig. 41, 4-hour on 2-hour off cycling) where mixing was turned off and particle settling was allowed. This "4-h resuspension_ON" and "2-h Resuspension_OFF" cycling was maintained over experiments of about 4 weeks duration (e.g., Fig. 42). Sediment resuspension was measured continuously in the STORM tanks using optical backscatter sensors (OBS-3) deployed 50 cm above the bottom and calibrated with direct total suspended solids (TSS) water samples taken during the course of the experiment.

Researchers performed four 4-week-long comparative ecosystem experiments with tidal resuspension using the STORM mesocosm facility (Fig. 40, Fig. 42). These experiments examined the effect of tidal resuspension on the ecosystem and on the contaminant and nutrient dynamics (Kim et al. 2004, 2006). In addition, experiments have varied benthic fauna in comparative ecosystem experiments with the STORM systems. Finally, experiments focusing on the effect of episodic sediment resuspension of Hudson River sediments on PCB release (Schneider et al. 2007) and particle dynamics were performed.

Physicochemical studies and contaminant transport studies that have been done include:

1. Examination of the partitioning of contaminants between the particulate phase and the dissolved phase, and the importance of *kinetic* (slow response time to changes in concentration) versus *equilibrium* (rapid attainment of steady state) control over the chemical distribution. These studies have shown that the notion of equilibrium partitioning between natural solids and dissolved constituents in coastal waters is not valid (Schneider 2005).

2. Examination of the rate of oxidation of sediments upon resuspension and the effect of such processes on metal and other contaminant

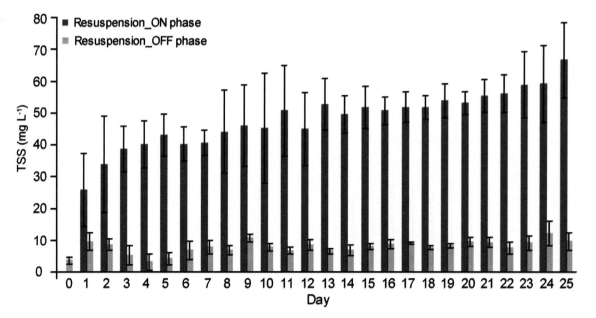

Figure 42: *Tidal resuspension (4-hour mixing on, 2-hour mixing off) maintained over an ~ 4-week period in the STORM systems during an ecosystem experiment, as measured using OBS-3 sensors. Representative data from one of the 4-hour mixing_ON phase and from the end of a 2-hour mixing_off phase are shown for each day of the experiment. Four of such mixing_on and mixing_OFF phases were programmed for each day for the duration of the experiment. The data show average TSS levels from three STORM tanks (± standard deviations). TSS = total suspended solids.*

release to solution. Mesocosm studies have found that the effect of resuspension is less than that obtained in smaller-scale studies where the energy for resuspension, and therefore the water column particulate load, is typically unrealistic (Kim et al. 2006). For methylmercury, the effect of resuspension on sediment mercury methylation rate can be effectively examined using STORM mesocosms because this process is dependent on physical, chemical, and biological factors in a complex fashion.

3. Studies of the effect of tidal resuspension and other physical processes on sediment chemistry and contaminant mobility across the sediment–water interface (Fig. 43). These have shown that resuspension has less effect on water-column metal concentrations than previously thought, although the effect is different for different metals. For example, cadium, which has a relatively low affinity for the solid phase, is bioaccumulated more strongly than metals that bind strongly with sediment, such as zinc and lead (Mason 2002; Langston et al. 1999; Schneider et al. 2007). The effect of resuspension on mercury dynamics is shown in Fig. 43.

Management implications
Mesocosm studies have shown that trace metals in particular, and by analogy other strongly-bound sediment contaminants, are not released to any significant degree by sediment resuspension. Release may occur during the initial resuspension of sediment but continual resuspension appears to result in decreased release to the water for PCB's, with the extent of release being a function of contaminant partitioning (Kim et al. 2006; Schneider 2005). This notion is consistent with the idea of the contaminant being distributed in both easily available and strongly bound pools in sediment. In addition, it appears that without continual input of the chemical from external sources, the fraction that remains in the easily available form decreases over time. These results reinforce the idea that contaminants that are strongly particle reactive are not readily bioavailable to aquatic organisms, and that the legacy of contamination in sediments may be less important than first expected.

However, because organic matter is often the major binding phase for these metals, and because organic content indirectly affects sediment redox state, changes in the ecosystem that result in a decrease in sediment organic carbon could lead to an increase in the release and availability of these contaminants to the food chain (Mason 2002; Schneider 2005; Di Torro et al. 2005; Bianchi 2007). Model results extrapolating the mesocosm results to the Chesapeake Bay ecosytem show the impact of sediment chemistry and resuspension on methylmercury bioaccumulation (Kim et al.

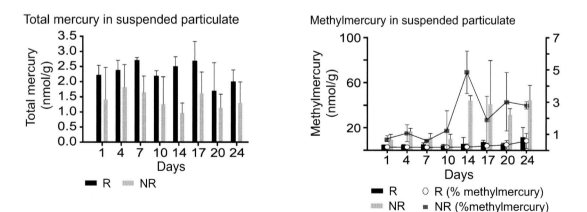

Figure 43: *Graphs showing the impact of sediment resuspension on the amount of total mercury and methylmercury in suspended particulates. While resuspension (R) increased the total mercury concentration, the percent methylmercury was higher for the non-resuspended (NR) mesocosms (Kim et al. 2005).*

2004, 2008; Fig. 44). The bioavailability of mercury and other metals both in the water and sediment to invertebrates and microbes is a strong inverse function of the organic content of the water or porewater. Thus, reductions in eutrophication may have a negative effect on contaminants by increasing bioavailability and bioaccumulation.

Finally, modeling studies have shown the importance of the rate of primary productivity in influencing the bioaccumulation of mercury, methylmercury, and likely other contaminants in food-limited environments (Di Torro et al. 2005; Kim et al. 2004, 2008; Ashley 1998). The implication is that food web structure and competition for resources are important considerations that are often not examined in sufficient detail when attempting to understand contaminant fate and bioaccumulation in coastal systems.

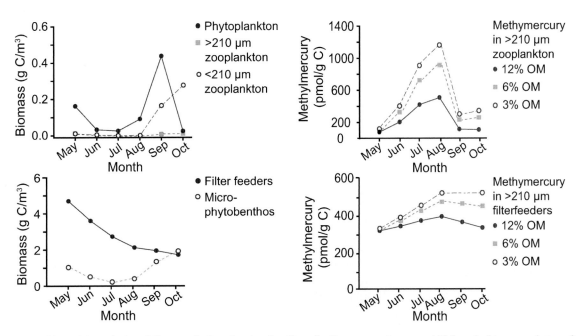

Figure 44: *Model simulation of Chesapeake Bay showing the effect of sediment organic content (OM) on the bioaccumulation of methylmercury into organisms. The model output also shows the biomass estimates for the different biota (Kim et al. 2004).*

Conclusions and recommendations
W.C. Dennison

The management applications provided in this book are representative of environmental problems facing coastal resource managers globally. The issues of aquatic grass declines, harmful algal blooms, biofiltration, salt marsh restoration, food web alteration and toxicants in the marine environment discussed in this book are all too common globally (Fig. 45). Excess nutrients leading to eutrophication is a global problem, with seagrass losses and proliferation of harmful algal blooms among the symptoms. For example, in a recent National Estuarine Eutrophication Assessment (Bricker et al. 2007), the majority of estuaries in the United States were highly influenced by human activities and seagrass losses and harmful algal blooms were two of the five indicators used to make this assessment. Global climate change and its

Jane Thomas

Figure 45: *Excess nutrients are impacting coastal systems across the globe. Knowledge gained from mesocosm experiments can help inform management decisions.*

influence on accelerated rates of sea-level rise has emerged as a serious problem for coastal habitats as well as coastal human populations. The importance of salt marshes in protecting coastal regions from storm surge has been recognized and large-scale restoration activities are being initiated in the Florida Everglades and coastal Louisiana, for example. Overfishing, both historic and modern, has led to food web alterations and loss of ecological services, such as biofiltration. Reestablishment of populations of biofiltering organisms is occurring for both commercial aquaculture as well as for their ecosystem services. The increased use of various toxicants in manufacturing, agriculture, and aquaculture leads to potential impacts in coastal ecosystems, and increasing knowledge about the fate and transport of toxicants needs to be developed. Thus, insights gained from mesocosm experiments that consider these issues have global ramifications.

Experimental ecosystem research can provide insight into causal mechanisms and aid in scaling up to management responses. Reports of global losses of coastal habitats such as seagrasses, mangroves, coral reefs and salt marshes provide an important indication that coastal ecosystems are deteriorating. In particular, global losses in mangrove forests and seagrass meadows provide a contrast in different causal factors. In the case of seagrasses, excess nutrients and sediments from runoff reduces the light reaching the seafloor and leads to plant losses. The case for mangroves (and salt marshes) is much different, in that excess nutrients and sediments stimulate their growth, but other factors like relative sea-level rise and shoreline hardening contribute to their demise. Since the different casual factors lead to very different management approaches, it is important to resolve the causal factors with a rigorous scientific approach. This is where experimental ecosystem research can make significant contributions to the application of the scientific results into society (Fig. 46). Simply documenting change in nature is often not sufficient to discern causal factors. In order to develop a management plan to alter aspects of human behavior, a clear causal link needs to be established. Management often involves enacting legislation, changing or enforcing rules, creating incentives, or expanding awareness. The scales of these management activities are much larger than the scales of traditional scientific experiments. The scaling of relationships established in experimental ecosystem research, thus, is an important part of the scientific application of results conducted at any scale.

Experimental ecosystem research addressing the issues of scale and complexity is crucial to building public confidence in management or research efforts and justifying large expenditures of public monies. Some of the largest public expenditures are motivated

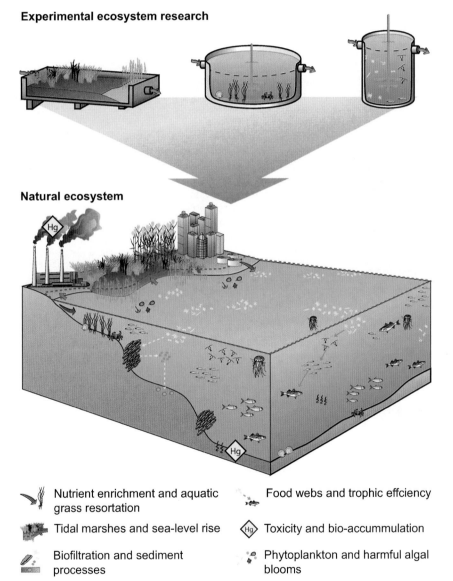

Experimental ecosystem research

Natural ecosystem

Nutrient enrichment and aquatic grass resortation

Tidal marshes and sea-level rise

Biofiltration and sediment processes

Food webs and trophic effciency

Toxicity and bio-accummulation

Phytoplankton and harmful algal blooms

Figure 46: Experimental ecosystem research render results that can be extrapolated to natural systems. Thus, this type of experiment can inform management decisions.

or influenced by environmental considerations including sewage treatment, reducing agricultural runoff, restoring ecosystem, and dredging. The public works aspect of these projects run into the billions of dollars for each region—in fact, costs for single sewage treatment facilities and dredging projects in Chesapeake Bay are on the order of a billion dollars each. To justify this large public expenditure of money, large scale "demonstration" projects are often required.

Table 1: *Scientific conclusions drawn from mesocosm experiments can lead to management recommendations for coastal ecosystems.*

Examples	Scientific conclusions	Management recommendations
Nutrient enrichment and aquatic grass restoration	Excess nutrients lead to aquatic grass declines, and restoration of aquatic grasses can be optimized by using existing aquatic grass communities as nurseries for transplanted grasses	Reduce nutrient loadings to coastal waters and optimize transplant success to enhance natural recovery
Tidal marshes, nutrient filtration, and rising sea level	Restored tidal marshes, even with low biodiversity, can effectively remove nutrients but marshes are increasingly threatened by rising sea levels	Augment existing tidal marshes with dredged sediments and restore degraded marshes
Biofiltration, water quality and sediment processes	Biofiltration by bivalves removes particles from the water column and leads to positive ecosystem outcomes (e.g., increased water clarity, stabilized sediments, reduced nutrient release by sediments)	Integrate increasing biofiltration capacity as part of ecosystem restoration
Food webs, trophic efficiency, and nutrient enrichment	Proposed nutrient reductions in coastal waters have beneficial outcomes of improved water quality and trophic efficiencies, without detrimental impacts on fisheries production	Reduce nutrient loadings into coastal waters, while monitoring attributes of the entire food web
Toxicity and bioaccumulation in benthic organisms	Food web dynamics and eutrophication status in coastal waters affect toxicant dynamics more than sediment resuspension	Reduce toxicant inputs into coastal waters and monitor bioaccumulation closely when restoration efforts change nutrient loadings
Phytoplankton, nutrients, and harmful algal blooms	Different forms of nutrients affect groups of phytoplankton differently, with organic nutrients promoting potentially harmful algal species	Reduce nutrient loadings of all forms of nutrients into coastal waters, with particular attention to organic nutrients

Photo: Adrian Jones
Photo: Heather Lane
Photo: NOAA
Photo: Adrian Jones
Photo: Joanna Woerner
Photo: Giancarlo Cichetti

References

Anderson, D.A., P.M. Glibert, and J.M. Burkholder. 2002. Harmful algal blooms and eutrophication: Nutrient sources, composition, and consequences. Estuaries 25: 562–584.

Ashley, J.F.A. 1998. Habitat use and trophic status as determinants of hydrophobic organic contaminant bioaccumulation within shallow systems. PhD Dissertation. University of Maryland, College Park, MD.

Bartleson, R.D., W.M. Kemp, and J.C. Stevenson. 2005. Use of a simulation model to examine effects of nutrient loading and grazing on *Potamogeton perfoliatus L.* communities in microcosms. Ecological Modelling. 185: 483–512.

Berg, G.M., P.M. Glibert, M.W. Lomas, and M. Burford. 1997. Organic nitrogen uptake and growth by the Chrysophyte *Aureococcus anophagefferens* during a brown tide event. Mar. Biol. 129: 377–387.

Berg, G.M., P.M. Glibert, and C.C. Chen. 1999. Dimension effects of enclosures on ecological processes in pelagic systems. Liminology and Oceanography 44: 1331–1340.

Bergeron, C.M. 2005. The impact of sediment resuspension on mercury cycling and the bioaccumulation of methylmercury into benthic and pelagic organisms. MS Thesis, University of Maryland, College Park, MD.

Berman, T., and D.A. Bronk. 2003. Dissolved organic nitrogen: A dynamic participant in aquatic ecosystems. Aquatic microbial ecology 31: 279–305.

Bianchi, T.S. 2007. Biogeochemistry of Estuaries. Oxford University Press, NY.

Borum, J. 1985. Development of epiphytic communities on eelgrass (*Zostera marina L.*) along a nutrient gradient in a Danish estuary. Marine Biology 87: 211–218.

Bricker, S., B. Longstaff, W. Dennison, A. Jones, K. Boicourt, C. Wicks, and J. Woerner. 2007. Effects of nutrient enrichment in the nation's estuaries: A decade of change. NOAA Coastal Ocean Program Decision Analysis Series No. 26. National Centers for Coastal Ocean Science, Sliver Spring, MD.

Bromilow, R.H., R.F. de Carvalho, A.A. Evans, and P.H. Nicholls. 2006. Behavior of pesticides in sediment/water systems in outdoor mesocosms. Journal of Environmental Science and Health. Part B 41: 1–16.

Brooks, M.T. 2004. Trophic complexity, transfer efficiency and microbial interactions in pelagic ecosystems: A modeling study. MS Thesis, University of Maryland, College Park, MD.

Caddy, J.F. 1993. Towards a comparative evaluation of human impact on fishery ecosystems of enclosed and semi-enclosed seas. Reviews in Fisheries Science 1: 57–95.

Cerco, C., and K. Moore. 2001. System-wide submerged aquatic vegetation model for Chesapeake Bay. Estuaries 24: 522–534.

Cosper, E.M., W.C. Dennison, E.J. Carpenter, V.M. Bricelj, J.G. Mitchell, S.H. Kuenstner, D. Colflesh, and M. Dewey. 1987. Recurrent and persistent brown tide blooms perturb coastal marine ecosystem. Estuaries 10(4): 284–290.

de Leiva Moreno, J.I., V.N. Agostini, J.F. Caddy, and F. Carocci. 2000. Is the pelagic-demersal ratio from fishery landings a useful proxy for nutrient availability? A preliminary data exploration for the semi-enclosed seas around Europe. ICES J. Mar. Sci. 57: 1091–1102.

Dennison, W.C., R.J. Orth, K.A. Moore, J.C. Stevenson, V. Carter, S. Kollar, P.W. Bergstrom, and R.A. Batiuk. 1993. Assessing water-quality with submersed aquatic vegetation. Bioscience 43(2): 86–94.

Di Torro, D.M., J.A. McGrath, D.J. Hansen, W.J. Berry, P.R. Paquin, R. Mathew, K.B. Wu, and R.C. Santore. 2005. Predicting sediment metal toxicity using a sediment biotic ligand model: Methodology and initial application. Environmental Toxicology and Chemistry. 24: 2410–2427.

Duarte, C. 1995. Submerged aquatic vegetation in relation to different nutrient regimes. Ophelia 41: 87–112.

Dugdale, R.C., and J.J. Goering. 1967. Uptake of new and regenerated forms of nitrogen in primary production. Liminology and Oceanography 12(2): 196–206.

EPA. 2000. Chesapeake Bay Program. Chesapeake 2000.

EPA. 2004. The incidence and severity of sediment contamination in surface waters of the United States. USEPA Office of Science and Technology, Washington, D.C, Report # EPA-823-R-04-007.

Gacia, E., and C. Duarte. 2001. Sediment retention by a Mediterranean *Posidonia oceanica* meadow: The balance between deposition and resuspension. Estuarine and Coastal Shelf Science. 52: 505–514.

Glibert, P.M., and D.G. Capone. 1993. Mineralization and assimilation in aquatic, sediment, and wetland systems. Pages 243–272 in R. Knowles and T.H. Blackburn (eds.). Nitrogen Isotope Techniques. Academic Press, San Diego, CA.

Glibert, P.M., and C. Heil. 2005. Use of urea fertilizers and the implications for increasing harmful algal blooms in the coastal zone. Pages 539–544 in Contributed papers, the 3rd International Nitrogen Conference, Science Press USA Inc.

Glibert, P.M., C.A. Heil, D. Hollander, M. Revilla, A. Hoare, J. Alexander, and S. Murasko. 2004. Evidence for dissolved organic nitrogen and phosphorus uptake during a cyanobacterial bloom in Florida Bay. Marine Ecological Progress Series. 280: 73–83.

Glibert, P.M., and C. Legrand. 2006. The diverse nutrient strategies of HABs: Focus on osmotrophy. Pages 163–176 in E. Graneli and J. Turner (eds.). Ecology of Harmful Algae. Springer, Heidelberg.

Glibert, P.M., S. Seitzinger, C.A. Heil, J.M. Burkholder, M.W. Parrow, L.A. Codispoti, and V. Kelly. 2005. The role of eutrophication in the global proliferation of harmful algal blooms: New perspectives and new approaches. Oceanography 18(2): 198–209.

Glibert, P.M., J. Harrison, C. Heil, and S. Seitzinger. 2006. Escalating worldwide use of urea – A global change contributing to coastal eutrophication. Biogeochemistry 77: 441–463.

Hagy, J.D. 2002. Eutrophcation, hypoxia and trophic transfer efficiency in Chesapeake Bay. PhD Thesis University of Maryland, College Park, MD.

Hallagraeff, G.M. 1993. A review of harmful algal blooms and their apparent global increase. Phycologia 32: 79–99.

Hengst, A.M. 2007. Restoration ecology of Potamogeton perfoliatus in mesohaline Chesapeake Bay: The nursery bed effect. MS Thesis, University of Maryland, College Park, MD.

Hillbricht-Illkowska, A. 1977. Trophic relations and energy flow in pelagic plankton. Polish Ecological Studies 3: 3–98.

Kemp, W.M. 2000. Seagrass ecology and management: An introduction. Pages 1–8 in S. Bortone (ed.). Seagrasses: Monitoring, Ecology, Physiology, and Management. CRC Publ., Boca Raton, FL.

Kemp, W.M., W.R. Boynton, J.C. Stevenson, R.R. Twilley, and J.C. Means. 1983. The decline of submerged vascular plants in upper Chesapeake Bay: Summary of results concerning possible causes. Marine Technical Society Journal. 17: 78–89.

Kemp, W.M., M.T. Brooks, and R.R. Hood. 2001. Nutrient enrichment, habitat variability and trophic transfer efficiency in simple models of pelagic ecosystems. Marine Ecology Progress Series. 223: 73–87.

Kemp, W.M., R. Batiuk, R. Bartleson, P. Bergstrom, V. Carter, G. Gallegos, W. Hunley, L. Karrh, E. Koch, J. Landwehr, K. Moore, L. Murray, M. Naylor, N. Rybicki, J.C. Stevenson, and D. Wilcox. 2004. Habitat requirements for submerged aquatic in Chesapeake Bay: Water quality, light regime, and physical-chemical factors. Estuaries 27: 363–377.

Kemp, W.M., W. Boynton, J. Adolf, D. Boesch, W. Boicourt, G. Brush, J. Cornwell, T. Fisher, P. Glibert, J. Hagy, L. Harding, E. Houde, D. Kimmel, W.D. Miller, R.I.E. Newell, M. Roman, E. Smith, and J.C. Stevenson. 2005. Eutrophication of Chesapeake Bay: Historical trends and ecological interactions. Marine Ecology Progress Series. 303: 1–29.

Kim, E.-H. 2004. The importance of physical mixing and sediment chemistry in mercury and methylmercury biogeochemical cycling and bioaccumulation within shallow estuaries. PhD Dissertation, University of Maryland, College Park, MD.

Kim, E.-H., R.P. Mason, E.T. Porter, and H.L. Soulen. 2004. The effect of resuspension on the fate of total mercury and methylmercury in a shallow estuarine ecosystem. Marine Chemistry 86: 121–137.

Kim, E.-H., R.P. Mason, E.T. Porter, and H.L. Soulen. 2006. The impact of resuspension on sediment mercury dynamics, and methylmercury production and fate: A mesocosm study. Marine Chemistry 102: 300–315.

Kim, E.-H., R.P. Mason, and C.M. Bergeron. 2008. A modeling study on methylmercury bioaccumulation and its controlling factors. Ecological Modeling. 218: 267-289.

Kirk, J.T.O. 1994. Light and Photosynthesis in Aquatic Ecosystems. Second Edition. Cambridge University Press, Cambridge, UK.

Landry, M.R. 1977. A review of important concepts in the trophic organization of pelagic ecosystems. Helgolander wis Meeresunters 30: 8–17.

Langston, W.J., G.R. Burt, and N.D. Pope. 1999. Bioavailability of metals in sediments of the Dogger Bank (central North Sea): A mesocosm study. Estuarine and Coastal Shelf Science. 48: 519–540.

Luo, J., and S.B. Brandt. 1993. Bay anchovy Anchoa mitchilli production and consumption in mid-Chesapeake Bay based on a bioenergetics model and acoustic measures of fish abundance. Marine Ecological Progress Series. 98: 223–236.

Madden, C.J., and W.M. Kemp. 1996. Ecosystem model of an estuarine submersed plant community: Calibration and simulation of eutrophication responses. Estuaries 19(2B): 457–474.

Madsen, K.N., P. Nilsson, and K. Sundback. 1993. The influence of benthic microalgae on the stability of a subtidal sediment. Journal of Experimental Marine Biology and Ecology. 170: 159–177.

Malone, T.C., D.J. Conley, P.M. Glibert, L.W. HardingJr., and K. Sellner. 1996. Scales of nutrient limited phytoplankton productivity: The Chesapeake Bay example. Estuaries 19: 371–385.

Malone, T.C., H.W. Ducklow, E.R. Peele, and S. Pike. 1991. Picoplankton carbon flux in Chesapeake Bay. Marine Ecology Progress Series 78: 11–22.

Mason, R.P. 2002. The bioaccumulation of mercury, methylmercury and other toxic elements into pelagic and benthic organisms. Pages 127–149 in M.C. Newman, M.H. Roberts, and R.C. Hale (eds.). Coastal and Estuarine Risk Assessment. CRC/Lewis Publ., Boca Raton, FL.

Melton, J.H. 2002. Environmental quality and restoration of mesohaline submerged aquatic vegetation. MS Thesis, University of Maryland, College Park, MD.

Naeem, S., J. Lindsey, P. Sharon, J.H. Lawton, and R.M. Woodfin. 1994. Declining biodiversity can alter performance of ecosystems. Nature 368: 734–737.

Naeem, S., K. Hakansson, J.H. Lawton, M.J. Crawley, and L.J. Thompson. 1996. Biodiversity and plant productivity in a model assemblage of plant species. Oikos 76: 259–264.

Nagel, J. 2007. Plant-sediment interactions and biogeochemical cycling for seagrass communities in Chesapeake and Florida Bays. PhD Thesis. University of Maryland, College, Park, MD.

Newell, R.I.E. 1988. Ecological changes in Chesapeake Bay: Are they the result of overharvesting the American oyster, Crassostrea virginica? Pages 536–546 in M.P. Lynch and E.C. Krome (eds.). Understanding the Estuary: Advances in Chesapeake Bay Research. Chesapeake Research Consortium Publication 129 (CBP/TRS 24/88), Gloucester, UK.

Nixon, S.W., and B.A. Buckley. 2002. "A strikingly rich zone" – Nutrient enrichment and secondary production in coastal marine ecosystems. Estuaries 25: 782–796.

Orihel, D.M., M.J. Paterson, C.C. Gilmour, R.A. Bodaly, P.J. Blanchfield, H. Hintelmann, R.C. Harris, and J.W.M. Rudd. 2006. Effect of loading rate on the fate of mercury in littoral mesocosms. Environmental Science and Technology. 40: 5992–6000.

Oviatt, C.A. 1994. Biological considerations in marine enclosure experiments: Challenges and revelations. Oceanography 7: 45–51.

Pauly, D., V. Christensen, J. Dalsgaard, R. Froese, and F. Torres. 1998. Fishing down the food chain. Science 279: 860–863.

Point, V.A., R.I.E. Newell, J.C. Cornwell, and M.S. Owens. 2002. Influence of simulated bivalve biodeposition and microphytobenthos on sediment nitrogen dynamics: A laboratory study. Liminology and Oceanography 47: 1367–1379.

Porter, E.T. 1999. Physical and biological scaling of benthic-pelagic coupling in experimental ecosystem studies. PhD Thesis, University of Maryland, College Park, MD.

Porter, E.T., J.C. Cornwell, L.P. Sanford, and R.I.E. Newell. 2004a. Effect of oysters Crassostrea virginica and bottom shear velocity on benthic-pelagic coupling and estuarine water quality. Marine Ecology Progress Series. 271: 61–75.

Porter, E.T., L.P. Sanford, G. Gust, and F.S. Porter. 2004b. Combined water-column mixing and benthic boundary-layer flow in mesocosms: Key for realistic benthic-pelagic coupling studies. Marine Ecology Progress Series 271: 43–60.

Romdhane, M.S., H.C. Eilertsen, O.K.D. Yahia, and M.N.D. Yahia. 1998. Toxic dinoflagellate blooms in Tunisian lagoons: Causes and consequences for aquaculture. Pages 80–83 in B. Reguera, J. Blance, M.L. Fernandez, and T. Wyatt (eds.). Harmful Algae. Xunta de Galicia and Intergovernmental Oceanographic Commission of United Nations Educational, Scientific and Cultural Organization, Paris, France.

Schneider, A.R. 2005. PCB desorption from resuspended Hudson River sediment. PhD Dissertation, University of Maryland, College Park, MD.

Schneider, A.R., E.T. Porter, and J.E. Baker. 2007. Polychlorinated biphenyl release from resuspended Hudson River sediment. Environmental Science Technology 41(4) 1097–1103.

Schulte, K. 2003. Spatial structure and heterogeneity in beds of the seagrass *Ruppia maritima* and comparison to ecological variables. MS Thesis, University of Maryland, College Park, MD.

Short, F.T., D. Burdick, and J.E. Kaldy. 1995. Mesocosm experiments quantify the effects of eutrophication on eelgrass, *Zostera marina*. Limnology and Oceanography 40: 740–749.

Short, F.T., and S. Wyllie-Echeverria. 1996. Natural and human-induced disturbance of seagrasses. Environmental Conservation 23: 17–27.

Smayda, T.J. 1997. Harmful algal blooms: Their ecophysiology and general relevance to phytoplankton blooms in the sea. Limonology and Oceanography 42: 1137–1153.

Stankelis, R.M., M. Naylor, and W.R. Boynton. 2003. Submerged aquatic vegetation in the mesohaline region of the Patuxent estuary: Past, present and future status. Estuaries 26(2A): 186–195.

Sturgis, R.B., and L. Murray. 1997. Scaling of nutrient inputs to submersed plant communities: Temporal and spatial variations. Marine Ecology Progress Series 152: 89–102.

Tilman, D. 1977. Resource competition between planktonic algae: An experimental and theoretical approach. Ecology 58: 338–348.

Tomasko, D.A., C.J. Dawes, and M.O. Hall. 1996. The effects of anthropogenic nutrient enrichment on Turtle grass (*Thalassia testudinum*) in Sarasota Bay, Florida. Estuaries 19(2B): 448–456.

Twilley, R.R., W.M. Kemp, K.W. Staver, J.C. Stevenson, and W.R. Boynton. 1985. Nutrient enrichment of estuarine submersed vascular plant communities: I. Algal growth and effects on production of plants and associated communities. Marine Ecology Progress Series 23: 179–191.

Yamamoto, T. 2003. The Seto Inland Sea – Eutrophic or oligotrophic? Marine Pollution Bulletin 47: 37–42.

Zedler, J.B., and J.C. Callaway. 1999. Tracking wetland restoration: Do mitigation sites follow desired trajectories? Restoration Ecology 7: 69–73.

Index

Printed in the United States of America